Vanadium oxides – properties and applications

KRYSTYNA SCHNEIDER

AMERICAN ACADEMIC PRESS

AMERICAN ACADEMIC PRESS

By AMERICAN ACADEMIC PRESS

201 Main Street

Salt Lake City

UT 84111 USA

Email manu@AcademicPress.us

Visit us at http://www.AcademicPress.us

ISBN: 979-8-3370-8926-3

Distributed to the trade by National Book Network Suite 200, 4501 Forbes Boulevard, Lanham, MD 20706

10   9   8   7   6   5   4   3   2   1

Contents

I. Introductory remarks..1

   1. INTRODUCTION..1

   2. VANADIUM OXIDE ELECTRONICS...1

   3. METAL-INSULATOR TRANSITIONS in VANADIUM OXIDES..................................2

   4. THIN FILMS – PREPARATION, PROPERTIES, APPLICATIONS................................2

   5. ELECTRONIC STRUCTURE...2

   6. DEFECT STRUCTURE and ELECTRICAL PROPERTIES of VANADIUM OXIDES......2

   7. CONCLUSIONS...2

   References...3

II. Vanadium oxide electronics 4

   1. INTRODUCTION..4

   2. VANADIUM-OXYGEN SYSTEM – PHASE DIAGRAM 4

   3. Thermodynamics of $V+O_2$ reactions 5

   4. SINGLE-VALENCE VANADIUM OXIDES 9

   5. MAGNÉLI AND WADSLEY HOMOLOGOUS VANADIUM OXIDES.........................16

   6. CONCLUSIONS...19

   References...19

III. Metal-insulator transitions in vanadium oxides...25

   1. INTRODUCTION..25

   2. MECHANISM OF A METAL-INSULATOR TRANSITION (MIT)..............................25

   3. MIT IN VANADIUM OXIDES..27

   4. MIT IN V_2O_3...29

   5. MIT IN VO_2..30

   6. MIT IN V_2O_5..31

   7. MIT – APPLICATIONS...38

   8. CONCLUSIONS...40

   REFERENCES..40

IV. Thin films – preparation, properties, applications..46

   1. INTRODUCTION..46

   2. PREPARATION OF VANADIUM OXIDE THIN FILMS – LITERATURE REVIEW 46

   3. PREPARATION OF THIN FILMS – OWN STUDIES...51

   4. CONCLUSIONS...60

   REFERENCES..60

V. Electronic structure...65

   1. INTRODUCTION..65

   2. ELECTRONIC STRUCTURE...65

   3. CONCLUSIONS...76

   REFERENCES..77

VI. Defect structure and electrical properties of vanadium oxides...81

   1. INTRODUCTION..81

2. DEFECT STRUCTURE OF V_2O_3..81
3. DEFECT STRUCTURE OF VO_2 THIN FILMS..83
4. ELECTRICAL PROPERTIES & DEFECT STRUCTURE OF V_2O_5 THIN FILMS....................86
5. CONCLUSIONS...102
REFERENCES..103

I. Introductory remarks

Abstract: This paper considers several issues related to the quantitative analysis of the properties of vanadium oxides and their applications. These issues will be addressed in more detail in follow-up papers in terms of the structural and defect chemistry properties as well as the related electronic, electrical and transport properties. This set of papers includes both a major collection of literature references and own unpublished data concerning the properties and potential practical applications of vanadium oxides as well as their critical analysis.

Keywords: vanadium oxide electronics, metal-insulator transition, thin films, electronic structure, point defect structure, chemical diffusion

1. INTRODUCTION

Solid metal oxides are the most abundant materials in Earth's crust. Their structure varies, and so do their chemical, mechanical, electrical, optical and magnetic properties. Metal oxide semiconductors are strikingly different from conventional inorganic semiconductors such as silicon and III-V compounds with respect to material design concepts, defect structure, electronic structure and charge transport mechanisms. This allows them to serve both conventional and completely new functions. Recently, remarkable advances in oxide electronics have been achieved. The term 'oxide electronics' had emerged not too long ago, but its place in the literature of the subject has already been firmly established [1].

The unprecedented diversity of physical properties exhibited by transition metal oxides offers immense prospects for various electronic applications. This is all the more significant given the fact that the modern IT revolution has been based on technological progress that has allowed the performance of electronic devices to grow at an exponential rate. Over the history of the development of electronic components ranging from a vacuum diode to modern highly integrated circuits with nanosized individual elements, the question of the impact of physical limitations on further progress in this area has become increasing relevant. The objective of the presented series of papers [2-6] was to analyse the available experimental and theoretical material on vanadium oxides. This introductory paper gives a short overview of the critical issues on this topic, which will be addressed more comprehensively in papers [2-6].

2. VANADIUM OXIDE ELECTRONICS

The outstanding physical and chemical properties of vanadium oxides give them an exceptional position among oxide materials [1, 7]. About 14 vanadium oxides are known. They can be classified as:
- single-valence vanadium oxides: VO, V_2O_3, VO_2 and V_2O_5
- double valence vanadium oxides: V_nO_{2n-1} ($2 < n < 10$) and V_nO_{2n+1} ($n = 3, 4, 6$) known as Magnéli and Wadsley series, respectively.

Reference [2] includes a vanadium-oxygen phase diagram, and it describes the thermodynamics of the formation of vanadium oxides as well as the thermodynamics of the partial reduction of vanadium pentoxide, which leads to the synthesis of respective lower valence vanadium oxides. Moreover, the crystallographic structure of vanadium oxides is discussed.

3. METAL-INSULATOR TRANSITIONS in VANADIUM OXIDES

One of the most spectacular phenomena observed for vanadium oxides is an abrupt change in electrical conductivity from one typical of an insulator or semiconductor to that typical of a metal phase. This phenomenon, called the metal-insulator transition (MIT), is observed for all vanadium oxide phases with the exception of VO and V_7O_{13}. Reference [3] describes a Mott-Hubbard metal-insulator transition. The temperature dependence of the metal-insulator transition, T_{MIT}, versus the chemical composition of vanadium oxide was analysed. A detailed description of the MIT in VO_2 and V_2O_5 and its practical applications is included.

4. THIN FILMS – PREPARATION, PROPERTIES, APPLICATIONS

Vanadium dioxide, VO_2, has been intensively studied due to its unique properties and the fact that it undergoes the metal-insulator transition (MIT) at near room temperature. It is currently considered one of the most promising materials for oxide electronics. Both planar and sandwich thin-film Metal-Oxide – Metal (MOM) devices based on VO_2 exhibit electrical switching with an S-shaped I-V characteristics, and this switching effect is associated with the MIT. In an electrical circuit containing such a switching device, relaxation oscillations are observed if the load line intersects the I–V curve at a unique point in the negative differential resistance (NDR) region [8]. Each of these effects is an advantage when designing various devices based on oxide electronics, especially the Mott-FET (field effect transistor based on the Mott MIT material) [9] or elements of dynamic neuron networks based on coupled oscillators [10]. Reference [4] describes the preparation, properties, and applications of vanadium oxides thin films.

5. ELECTRONIC STRUCTURE

Reference [5] describe the electronic properties of vanadium oxides. The electronic structure of the three main vanadium oxides – V_2O_3, VO_2 and V_2O_5 – is reviewed. The electronic properties were studied via optical measurements. Optical transmittance and reflectance spectra were measured over a wide wavelength range with a Lambda 19 Perkin-Elmer double beam spectrophotometer equipped with a 150 mm integrating sphere. The optical properties such as energy band gap, of vanadium pentoxide thin films were determined. It was found that a direct allowed (DA) transition is the most probable one in the studied films.

6. DEFECT STRUCTURE and ELECTRICAL PROPERTIES of VANADIUM OXIDES

The electrical properties of vanadium oxides were investigated by analyzing the complex impedance spectra (IS) as a function of temperature (T), oxygen partial pressure ($p(O_2)$) and equilibration time (t) of electrical conductivity, σ. Based on the dependence $\sigma = f[p(O_2)]$, the predominant type of point defects was determined. The dependence $\sigma = f(T)$ provided values of the activation energy of conductivity and the corresponding enthalpy of point defect formation. On the other hand, $\sigma = f(t)$ made it possible to determine the chemical diffusivity of point defects.

7. CONCLUSIONS

The physicochemical properties of vanadium oxides have been the subject of many conflicting reports. The presented series of papers [2-7] provides an extensive analysis of the available experimental data on the their structure and the related properties, in particular, the electrical properties that determine the feasibility of the practical application of vanadium oxides as materials used for the production of oxide electronics.

References

1. *Oxide electronics and functional properties of transition metal oxides*, A. Pergament, ed., New York: NOVA Sci. Publishers, 2014.
2. K. Schneider, *Vanadium oxides – properties and applications*. II *Vanadium oxide electronics,* this issue.
3. K. Schneider, *Vanadium oxides – properties and applications*. III. *Metal-Insulator Transition (MIT), vanadium oxides,* this issue.
4. K. Schneider, *Vanadium oxides – properties and applications*. IV. *Thin films: preparation, properties, applications,* this issue.
5. K. Schneider, *Vanadium oxides – properties and applications*. V. *Electronic structure,* this issue.
6. K. Schneider, *Vanadium oxides – properties and applications*. VI *Defect structure and electrical properties,* this issue
7. V.E. Henrich, P.A. Cox, *The surface science of metal oxides*, University Press, Cambridge, 1994.
8. A. Pergament, G. Stefanovich, and A. Velichko, *Oxide electronics and vanadium dioxide perspective: A review,* J. Sel. Top. Nano Electron. Comput. 1, 24 (2013).
9. Z. Yang, C. Ko, and S. Ramantham, *Oxide electronics utilizing ultrafast metal-insulator transitions,* Ann. Rev. Mater. Res. 41(2011) 337-367.
10. A. Beaumont, J. Leroy, J.-C. Orlianges, and A. Crunteanu, *Current-induced electrical self-oscillations across out-of-plane threshold switches based on* VO_2 *layers integrated in crossbars geometry,* J. Appl. Phys. 115, 154502 (2014).

II. Vanadium oxide electronics

Abstract

Vanadium oxides can exist as single- and mixed-valence compounds with a large variety of structures. They exhibit diverse physicochemical properties which make them important with regard to oxide electronic material technology. This paper presents a broad overview of single-valence vanadium oxides and the mixed-valence variants, which form the Magnéli and Wadsley homologous series. Under certain ambient conditions (including temperature), phase transformations between these oxides can occur. Based on the available thermodynamic data, the specific conditions required to obtain particular oxides were determined.

Keywords: vanadium oxides, single-valence oxides, Magnéli phases, Wadsley phases thermodynamics, crystal structure

1. INTRODUCTION

Vanadium belongs to the transition metal group with the electronic configuration $[Ar]3d^34s^2$, which means that in compounds this element can assume the valence of +2 (V^{2+} with the $[Ar]3d^3$ electron configuration), +3 ($V^{3+}[Ar]3d^2$), +4 ($V^{4+}[Ar]3d^1$) and +5 ($V^{5+}[Ar]$). There is a large number of vanadium oxide phases that take the form of either single-valence binary oxides (VO, V_2O_3, VO_2, V_2O_5) or oxides with two different oxidation states of vanadium, known as Magnéli or Wadsley phases.

2. VANADIUM-OXYGEN SYSTEM – PHASE DIAGRAM

The phase diagram of vanadium-oxide was the subject of several papers [1-4]. Based on both these data and the newest results on the melting temperatures of vanadium oxides, the phase diagram for an oxygen-to-vanadium ratio corresponding to 40-72 at.% of O was constructed (Fig. 1).

Little is known about the phase diagram for low oxygen content (below 40 at.%). Metal vanadium may dissolve relatively large amounts of oxygen (up to 3 at.% at 1770 K) [5].

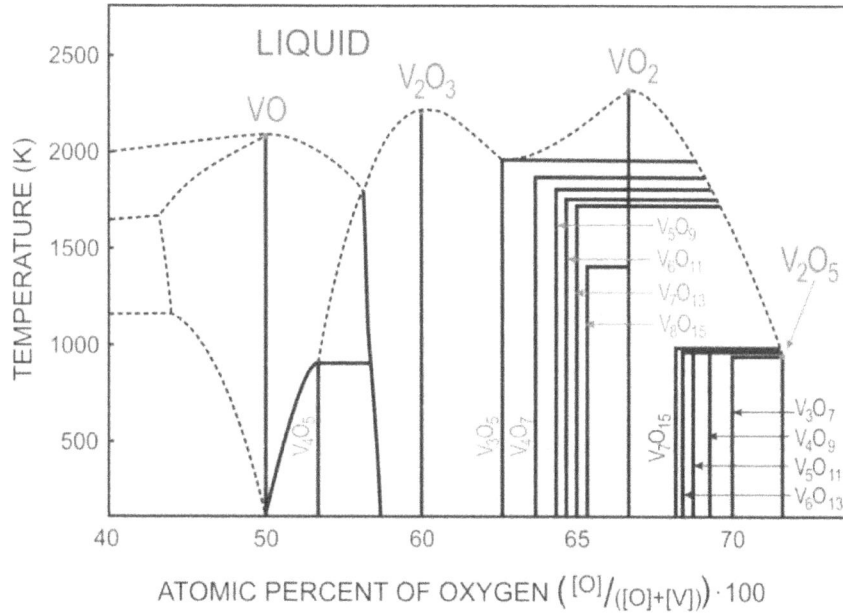

Fig.1 Vanadium-oxygen phase diagram. Data compiled from papers [1-4].

3. Thermodynamics of V+O₂ reactions

3.1 Definition of terms

Thermodynamics is a branch of physics and chemistry concerned with the interdependences between heat, work, temperature and energy in physical processes and chemical reactions. Knowledge of the energy transfer that takes place during physical and chemical transitions makes it possible to predict the type of changes that are likely to occur. The general rule is based on the second law of thermodynamics: the physical-chemical processes will (or can) proceed spontaneously if the change in the total entropy (S) of the universe resulting from this process is non-negative. In the case of chemical reactions that occur at a constant temperature (T) and pressure (p), this rule assumes the following form: at constant T and p, the Gibbs free energy (G) attains a minimum. From the rule which defines the change in the Gibbs free energy for a chemical reaction as:

$$\Delta G = G_{products} - G_{reactants} \tag{1}$$

the following criteria can be formulated:

- if $\Delta G < 0$, then the reaction is possible
- if $\Delta G = 0$, the system is in equilibrium
- if $\Delta G > 0$, the reaction is impossible.

In particular, Eq. (1) can be applied to the reaction of the formation of the V_mO_n vanadium oxide:

$$mV + \frac{n}{2}O_2 \mapsto V_mO_n \tag{2}$$

$$\Delta G = \mu^o(V_mO_n) - m\mu^o(V) - \frac{n}{2}\left[\mu^o(O_2) + RT\ln p_{O2}\right] \tag{3}$$

where $\mu^o = G^o(A)$ [J/mole] is the chemical potential of pure component A, R is the gas constant $R = 8.3144$ J/(mole·K), T represents temperature [K], and p_{O2} [atm.] is oxygen activity, which can be expressed as oxygen partial pressure.

Based on the first and third laws of thermodynamics as well as on the experimentally determined temperature dependence of specific heat, the values of chemical potential vs. temperature were determined for various phases, including gases, liquids and solids of elements and chemical compounds.

3.2 Thermodynamics of vanadium oxide formation

5

When the ambient gas atmosphere exhibits certain characteristics, i.e. oxygen activity, p_{O2}, and temperature (T), phase transformations between vanadium oxides can occur. These conditions can be estimated via the thermodynamic calculation of the changes in Gibbs energy (ΔG) that accompany chemical reactions. The most comprehensive thermodynamic data on chemical potentials (μ^o) for inorganic substances are collected in [1]. In the case of the vanadium-oxygen system, the available thermodynamic data include those for V and O_2, those for single-valence vanadium oxides (VO, V_2O_3, VO_2 and V_2O_5), and those for oxides that belong to the Magnéli series – V_3O_5 and V_4O_7 [1].

Fig. 2 illustrates the standard Gibbs energy (G^o corresponding to $p_{O2} = 1$ atm.) of the formation of principal (single-valence) vanadium oxides as a function of temperature.

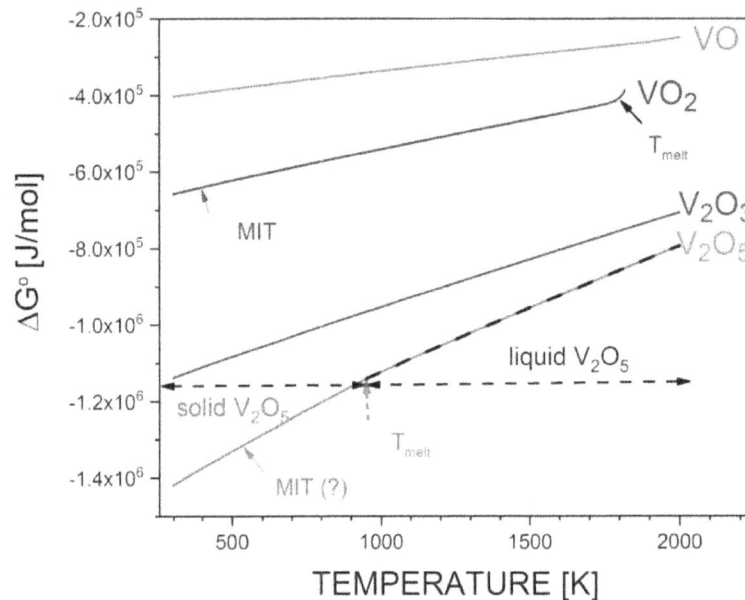

Fig.2 Standard Gibbs energy ($p_{O2} = 1$ atm.) of the formation of principal vanadium oxides.

The ΔG^o assumes negative values within the whole stability range of all oxides. This means that the oxidation of metal vanadium under standard conditions, i.e. $p_{O2} = 1$ atm., can occur spontaneously. The lowest ΔG^o is observed in the case of the formation of V_2O_5. Fig. 3 shows the temperature dependence of ΔG^o for all available thermodynamic data in the range of 298-1000 K, i.e. the range corresponding to the chemical stability of mixed-valence oxides V_3O_5 and V_4O_7.

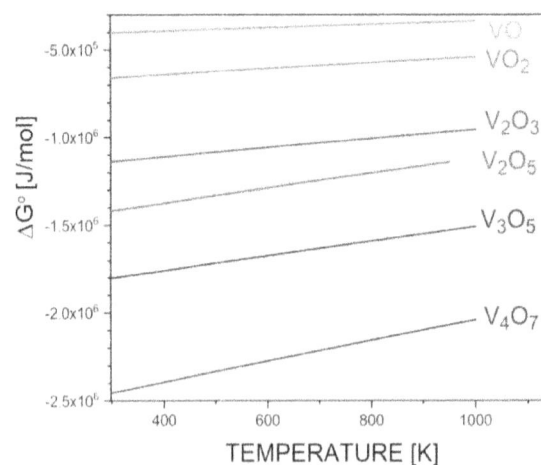

Fig.3 Standard Gibbs energy ($p_{O2} = 1$ atm.) of the formation of vanadium oxides in the range of 298-1000 K.

In order to determine which oxide will be the final product of the oxidation under such conditions, the values of G^o corresponding to 1 mole of vanadium need to be compared. For instance, in the case of V_4O_7, the standard Gibbs energy

(ΔG^{o}) of vanadium formation [J/mole], can be determined from Eqs (2) and (3):

$$4V + \frac{7}{2}O_2 \Rrightarrow V_4O_7 \tag{4}$$

$$\Delta G^{o}(V_4O_7) = \mu^{o}(V_4O_7) - 4\mu^{o}(V) - \frac{7}{2}[\mu^{o}(O_2) \ [J/mol] \tag{5}$$

which for 1 mole of vanadium gives:

$$V + \frac{7}{8}O_2 \Rrightarrow \frac{1}{4}V_4O_7 \tag{6}$$

$$\Delta G^{o}_{(1 \ mol \ V)} = \frac{1}{4}\mu^{o}(V_4O_7) - \mu^{o}(V) - \frac{7}{8}[\mu^{o}(O_2) = \frac{1}{4}\Delta G^{o} \ (V_4O_7) \tag{7}$$

Such dependence for 1 mole of V is presented in Fig. 4.

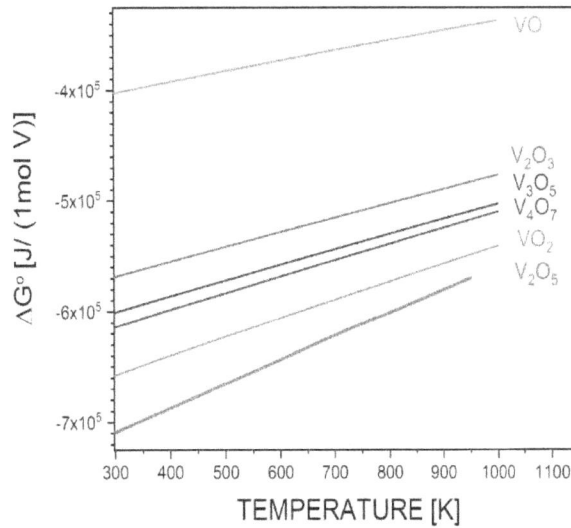

Fig.4 Standard Gibbs energy ($p_{O2} = 1$ atm.) corresponding to the oxidation of 1 mole of vanadium.

Vanadium pentoxide exhibits the lowest negative value of ΔG^{o} [J/(1 mole of V)], which indicates that V_2O_5 is the most stable phase. By applying Eq. (3), the change in the Gibbs energy (ΔG) can be determined as a function of T and p_{O2}. Fig. 5 illustrates the ΔG of the formation of vanadium pentoxide as a function of temperature for several p_{O2} values. As the presented figure shows, vanadium oxidation is a spontaneous process for the oxygen partial pressure range of 1-10^{-15} atm. Outside this range – i.e. at 10^{-20} and 10^{-30} atm. – vanadium oxidation cannot occur at temperatures above 840 and 1130 K, respectively. In other words, under such conditions vanadium pentoxide can spontaneously reduce to metallic vanadium. Such low oxygen activity can be achieved experimentally using a gas buffer composed of an H_2/H_2O gas mixture.

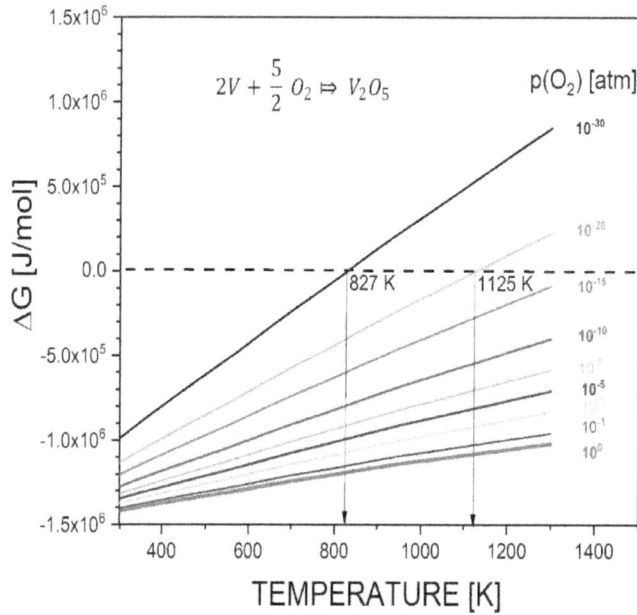

$$2V + \frac{5}{2}O_2 \rightleftharpoons V_2O_5$$

Fig.5 Gibbs energy of the formation of vanadium pentoxide for several oxygen partial pressures (p_{O2}).

3.3 Thermodynamics of partial reduction of vanadium pentoxide

Vanadium pentoxide is typically used as a starting reactant in the preparation of lower vanadium oxides via the partial reduction of V_2O_5. The condition under which spontaneous partial reduction occurs can be determined from the suitable reaction equations (Eq. (2) and Eq. (3)).

The conditions for the reduction of V_2O_5 to vanadium dioxide are shown in Fig. 6. The process can proceed simultaneously ($\Delta G < 0$) below 1000 K if p_{O2} assumes a value of 10^{-5} atm. or lower. Such conditions can easily be achieved using an H_2/Ar gas mixture [6]. The conditions that allow the reduction of vanadium pentoxide to vanadium sesquioxide (V_2O_3) can be established from Fig.7.

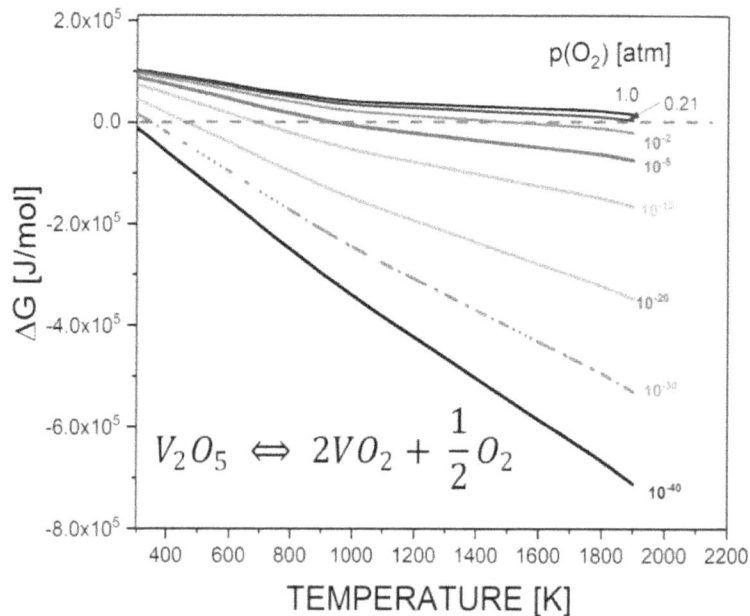

$$V_2O_5 \Longleftrightarrow 2VO_2 + \frac{1}{2}O_2$$

Fig.6 Gibbs energy of the partial reduction of vanadium pentoxide to vanadium dioxide at several oxygen partial pressures (p_{O2}).

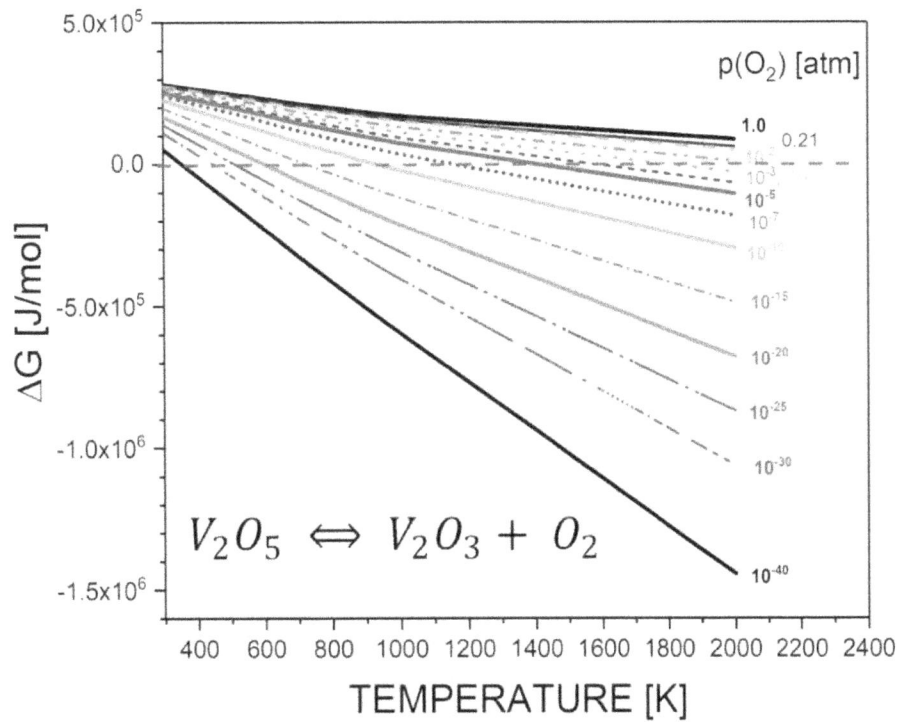

Fig.7 Gibbs energy of the partial reduction of vanadium pentoxide to vanadium sesquioxide for several oxygen partial pressures (p_{O2}).

4. SINGLE-VALENCE VANADIUM OXIDES

The principal oxides of vanadium in single-valence form ranging from V^{2+} to V^{5+} are VO, V_2O_3, VO_2 and V_2O_5. The basic properties of vanadium and principal vanadium oxides are shown in Table 1.

Table 1. Basic properties of vanadium and principal (single-valence) oxides of vanadium

Material formula/name	Vanadium valence	T_{MIT} [K]	Low T Structure, E_g ($T < T_{MIT}$)	High T Structure, E_g ($T > T_{MIT}$)	Melting point [7] [K]
V Vanadium	0	-	Body-centered cubic (bcc); $a = 303$ pm [8]; $E_g = 0$ eV		2183
VO Vanadium monoxide	2	-	Cubic, NaCl-type structure; $a = 404.6$ pm [9]		2062
V_2O_3 Vanadium sesquioxide	3	155 [10]	Monoclinic, I2/a (C^6_{2h}), karelianite structure; $T = 148$ K: $a = 725.5$ pm, $b = 500.2$ pm, $c = 554.8$ pm; $\beta = 96.75°$ [11]	Trigonal corundum, R3.ch space group (D^6_{3d}); $E_g = 0.66$ eV; $T = 298$ K: $a = 495.17$ pm, $c = 1400.5$ pm [12]	2210

9

VO₂ Vanadium dioxide	4	340 [13]	Monoclinic; P2₁/c space group (C⁵₂ₕ); at room temperature, VO₂ has a distorted rutile structure with shorter distances between pairs of V atoms indicating metal-metal bonding; $T = 298$ K: $a = 575.173$ pm, $b = 45596$ pm, $c = 538.326$ pm; $\beta = 126148°$ [14]; $E_g = 0.6(5)$ eV [15]	Rutile, P4₂/mnm space group (D¹⁴₄ₕ,); a\360 K: $a = 455.396$ pm, $c = 285.028$pm [14]	2240
V₂O₅ Vanadium pentoxide	5	530 [16]	Orthorhombic Pnmn (D¹³₂ₕ); $a = 1150(8)$ pm, $c = 356.3(3)$ pm, $b = 436.9(5)$ pm [17]; H.G. $E_g = 23$ eV $E\|a$, 25 $E\|c$ [18] 19 eV [19]		963

4.1 VO

Like many other monoxides of transition metals, VO has a sodium-chloride-type structure (Fig.8). It is stable for a wide range of non-stoichiometric compositions, which range from about VO₀.₈₅ to VO₂₅. Within the oxygen deficit region ($x <$ 1), VO$_x$ contains a high concentration of both oxygen and vanadium vacancies. On the other hand, for $x > 1$, the concentration of anion vacancies is negligibly low [20]. The lattice parameter increases with increasing x.

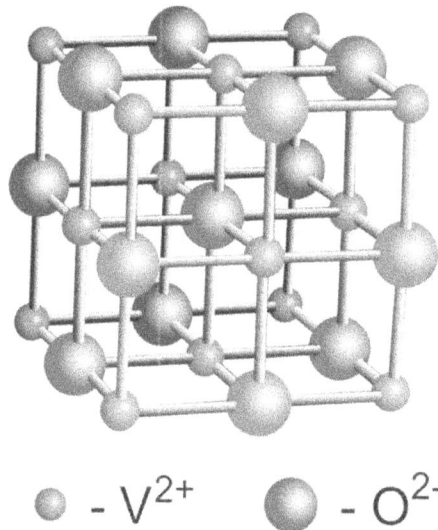

Fig.8 Crystallographic unit cell of VO.

The electrical conductivity of VO$_x$ decreases with increasing x, and the corresponding activation energy increases with x. It is used as a vanadium source suitable for applications such as glass and ceramics.

4.2 V₂O₃

Vanadium sesquioxide (V₂O₃) is one of the most extensively studied materials with a Mott-Hubbard metal-insulator transition. V₂O₃ is an antiferromagnetic material that undergoes the metal insulator transition (MIT) at $T_{MIT} = 150$ K [21], at which point its electrical conductivity changes by a factor of 10^5-10^6. The transition is accompanied by a transformation from a monoclinic structure (below T_{MIT}) to a corundum structure (above T_{MIT}). The monoclinic structure contains 20 atoms in its unit cell ($z = 4$; z – number of molecules), while the corundum primitive unit cell contains 10 atoms. McWahan and Rice [22] found the metal-semiconductor transition in V₂O₃ to shift by ca. 2 K at pressures above 6·10⁹ Pa. Doping with Cr has a marked effect on the lattice parameters and the MIT. Cr doping causes the MIT to occur at room temperature

[23]. Vanadium sesquioxide has a metal deficit ($V_{2-y}O_3$). The oxygen pressure dependence over the range of 1400-1700 K of y non-stoichiometry can be expressed as $y \propto p_{O2}^{3/4}$ [24]. This dependence may be represented as follows:

$$\frac{3}{2}O_2 \leftrightarrow 2V_{Va}^x + 3O_O \tag{8}$$

where V_{Va}^x denotes neutral vanadium vacancies (according to the Kröger-Vink notation).

The Kröger-Vink point defect notation is employed in this equation with one modification. According to Kröger-Vink notation, the symbol for vacancy should be V, but this is also the atomic symbol of vanadium; the symbol V_V would thus be ambiguous and represent either a vanadium atom in the vanadium site, or a vanadium vacancy. In order to avoid confusion, the symbol for vanadium vacancy used in Eq. (8) and throughout the present work is V_{Va}. The application of the law of mass action yields:

$$y = [V_{va}^x] = K_V^{1/2} \, p_{O2}^{\frac{3}{4}} \tag{9}$$

where K_V is the equilibrium constant of reaction (8). Based on the thermogravimetric results reported by Wakihara and Katsura [24], the equilibrium constant K_V may be determined from Eq. (9):

$$K_V = 3.185 \cdot 10^{-11} \exp\frac{546.8 \text{ kJ/mol}}{RT} \tag{9a}$$

Fig. 9 illustrates the dependence of the non-stoichiometry in $V_{2(1-y)}O_3$ on temperature (T) and oxygen partial pressure (p_{O2}). At a constant oxygen pressure, non-stoichiometry decreases with increasing temperature. This means that the enthalpy of vanadium vacancy formation is negative. Assuming that Eqs (8) and (9) describe the defect structure, the electroneutrality condition is given by:

$$n = p \tag{9b}$$

where n and p denote the concentration of electrons and electron holes, respectively.

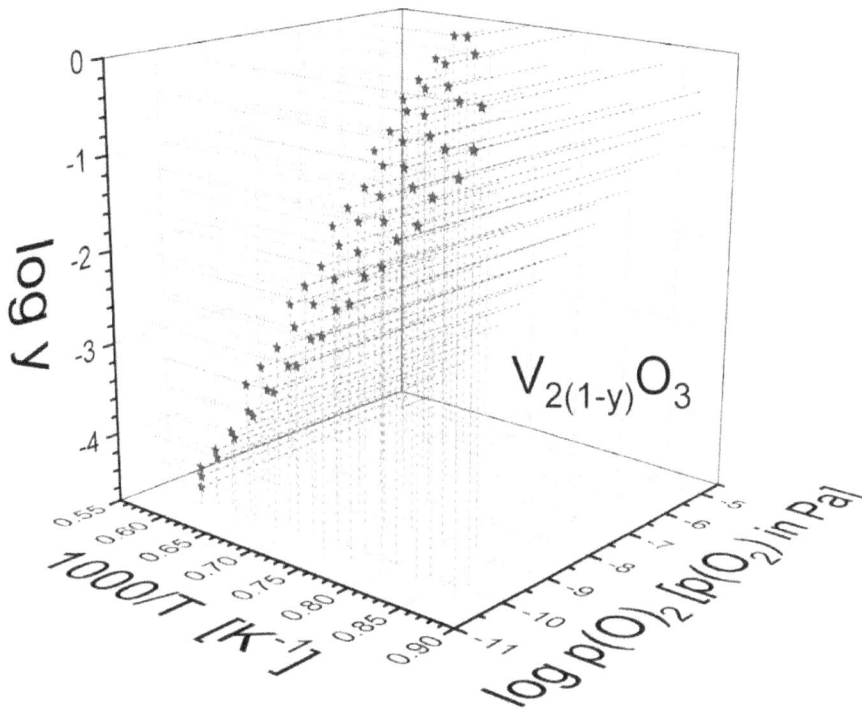

Fig.9 Deviation from stoichiometry in $V_{(2-y)}O_3$ as a function of temperature (T) and oxygen partial pressure (p_{O2}). V_2O_3 can be obtained by reducing V_2O_5 by means of hydrogen or carbon monoxide [25].

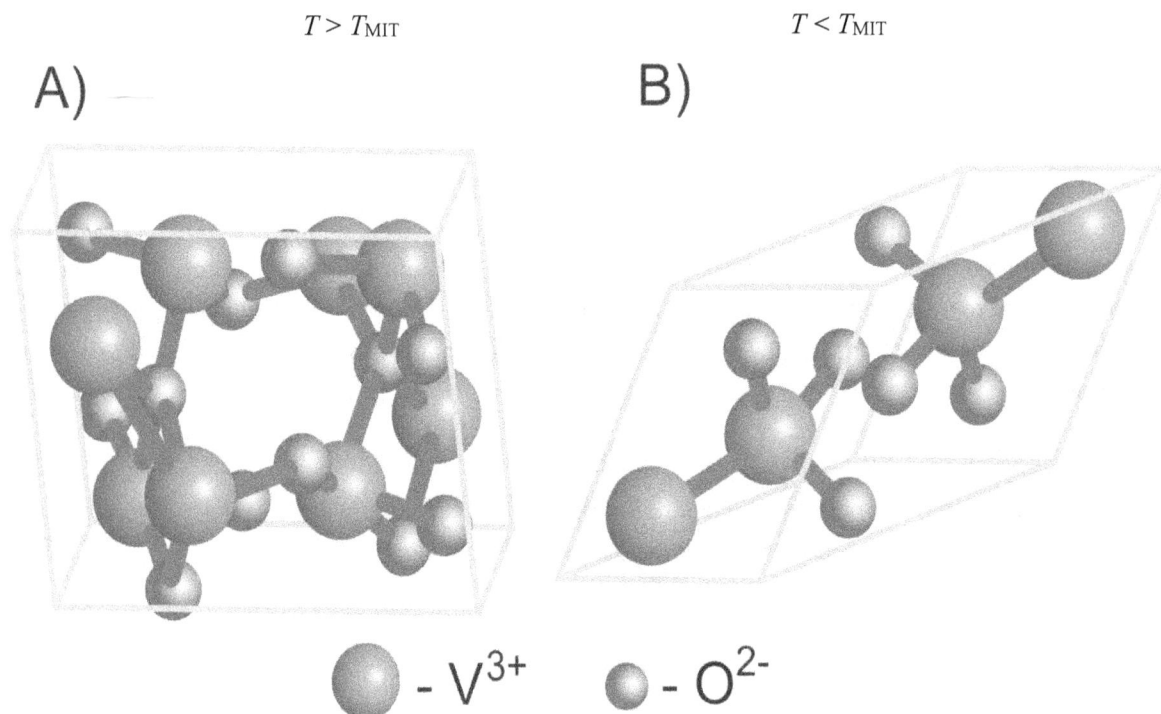

Fig.10 Crystallographic unit cells of the V_2O_3: A – trigonal (above T_{MIT}), B – monoclinic (below T_{MIT}).

Doping with a small amount of Al, Cr, W or Ti can change the temperature at which V_2O_3 undergoes the metal-insulator transition. This reversible structural transition makes V_2O_3 a very attractive candidate for a wide variety of technological applications. The material shows pronounced thermochromic behaviour in the infrared while retaining transparency in the visible spectral range [26].

Different methods have been used to synthesize V_2O_3 nanocrystals, including thermal reduction, thermal decomposition and hydrothermal synthesis [27].

V_2O_3 nanomaterial is important due to its potential for application in nano-optical /electronic devices. Luo et al. investigated the dynamic behaviour of thermally-induced metal-insulator transition in V_2O_3 thin films by means of tetrahertz time-domain spectroscopy. They found that the THz absorption and optical conductivity of the thin films they studied were temperature-dependent, and the THz amplitude modulation would reach a value as high as 75% [28]. Y. Guo and J. Robertson [29] proposed the use of V_2O_3 for the manufacture of resistive random-access memory (RRAMs). S. Long et al. [30] studied the thermochromic properties of a V_2O_3/VO_2 bilayer structure. V_2O_3 acts as a buffer layer, improving the crystallinity and durability of VO_2.

4.3 VO_2

Vanadium dioxide is currently considered one of the most promising materials for oxide electronics such as terahertz materials, field-effect devices, memory devices and sensors [31-33]; this is because its behaviour changes abruptly from semiconducting to metallic at 240-243 K. This change is associated with a crystallographic transition from a monoclinic structure at low temperatures to the tetragonal (rutile) structure [34]. Detailed crystallographic data for vanadium dioxide are listed in Table 2 [14, 35-42] and presented schematically in Fig. 10.

Vanadium dioxide has several polymorphs, including both stable phases VO_2 (M) and VO_2 (R) and metastable phases VO_2 (B), VO_2 (A_L) and VO_2 (A_H).

Anderson [40] synthesized single crystals of VO_2 by heating a mixture of V_2O_3 and V_2O_5 at 1173 K. The crystals developed with a monoclinic structure, and are marked VO_2 (M) in Table 2.

Westman [41] prepared polycrystalline vanadium oxide by heating the V_2O_3/V_2O_5 mixture at 1173 K for 20 days. The material had a rutile-type tetragonal structure – VO_2 (R).

Of the afore-mentioned vanadium dioxide polymorphs, VO₂ (M) and VO₂ (R) have been the most widely studied phases on account of their metal-insulator transition (MIT) temperature being close to room temperature. The remaining VO₂ polymorphs are metastable phases. Their physical properties and potential for technical applications have not been explored extensively. Various preparation techniques have been used to stabilize these phases in bulk and thin film forms [35, 39].

Oka et al. [42] reported the stabilization of low-temperature VO₂ (A$_L$) and high-temperature VO₂ (A$_H$) using hydrothermal synthesis. These phases exhibit a weak MIT at 435 K. On the other hand, the monoclinic VO₂ (B) does not undergo the metal-insulator transition [37].

Table 2. Crystallographic features of polymorphous VO₂ phases

Phase	T_{MIT} [K]	Crystal system	Space group	Lattice parameters				Ref.
				a [nm]	b [nm]	c [nm]	β [°]	
VO₂ (B)	-	monoclinic	C2/m	2093	0.3702	0.6433	106.97	35
				203	0.369	0.642	106.6	37
VO₂ (M)	341	monoclinic	P2₁/c	0.5743	0.4517	0.5375	126.1	36, 40
				0.538	0.452	0.574	126	37
				0.5752	0.4526	0.5383	126.15	14
VO₂ (R)	341	tetragonal	P4₂/mnm	0.4530	0.4530	0.2869	90.00	41
				0.45540	0.45540	0.28503	90.00	14
VO₂ (A$_L$)	435	tetragonal	P4/ncc	0.8440	0.8440	0.7666	90.00	35
				0.843	0.843	0.768	90.00	38
VO₂ (A$_H$)	435	tetragonal	I4/m	0.8476	0.8476	0.3824	90.0	35

Both planar VO₂ and sandwich thin-film metal/oxide/metal devices based on VO₂ exhibit electrical switching with an S-shaped I-V characteristic (Fig.11), and this switching effect is associated with the metal-insulator transition (MIT). In an electrical circuit containing such a switching device, relaxation oscillations are observed if the load line intersects the I-V curve at a unique point in the negative differential resistance region.

The remarkable first order insulator-to-metal phase transition in VO₂ has been studied for more than half a century. In addition to a fundamental interest in the nature of this phase transition, a variety of applications have been proposed and, for the most part, experimentally tested, as described in papers; examples of patents include the following: resistive switching elements, thermal relays, optical storage devices, holographic recording media, variable reflectivity mirrors, light modulators, energy-efficient windows, flat panel displays, and others.

It was found that point defects strongly affect the properties of monoclinic VO₂ [43-47].

Kim et al. [47] observed a decrease in the T_{MIT} when VO₂ films grew under oxygen deficient conditions. Yu et al. [47] found that the hysteresis curve across the phase transition grew narrower at a high oxygen pressure. However, what type of point defects is formed at a low or high oxygen partial pressure is still an open question [47-51]. Some authors suggests that oxygen and vanadium vacancies are the predominant defects in the case of an oxygen deficit and excess, respectively. Others propose that the corresponding defects are oxygen vacancies and oxygen interstitials.

A)

B)

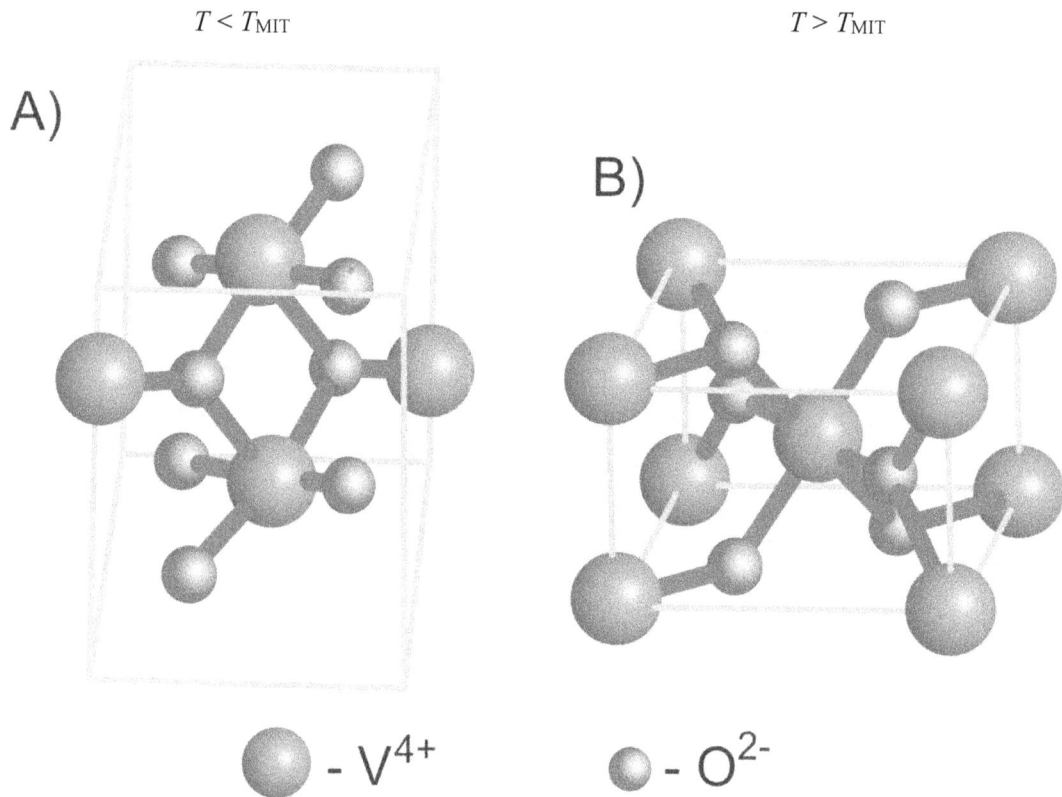

$\bigcirc$ - V^{4+} $\circ$ - O^{2-}

Fig.11 Structures of VO₂: A) monoclinic, B) tetragonal (rutile).

4.4 V₂O₅

Vanadium pentoxide (V_2O_5) exhibits the highest oxidation state of vanadium (+5), with the electron configuration of [Ar]; it is therefore the most stable compound in the V-O system. It exhibits an orthorhombic structure belonging to the P_{mnm} space group [52] (Fig. 12).

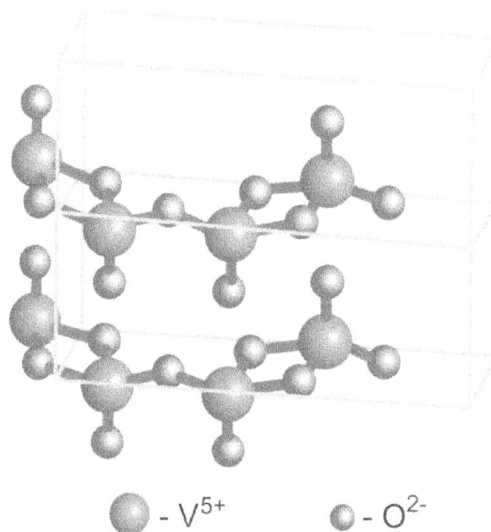

$\bigcirc$ - V^{5+} $\circ$ - O^{2-}

Fig.12 Crystallographic unit cell of V₂O₅.

It has a layered structure and it is composed of distorted trigonal, bi-pyramidal coordination polyhedra of O atoms arranged around V atoms. The polyhedra share edges, forming ($V_2O_4)_n$ zigzag double chains along the (001) direction, and are cross-linked along (100) through shared corners, as shown in Fig. 13. The distorted polyhedra have a short (0.158 nm) vanadyl bond, $[VO]^{2+}$, and four O atoms located in the basal plane at distances ranging from 0.178 to 0.202 nm. The sixth O atom in the coordination polyhedron lies along the vertical axis opposite to the V-O bond at a distance of 0.279 nm [53].

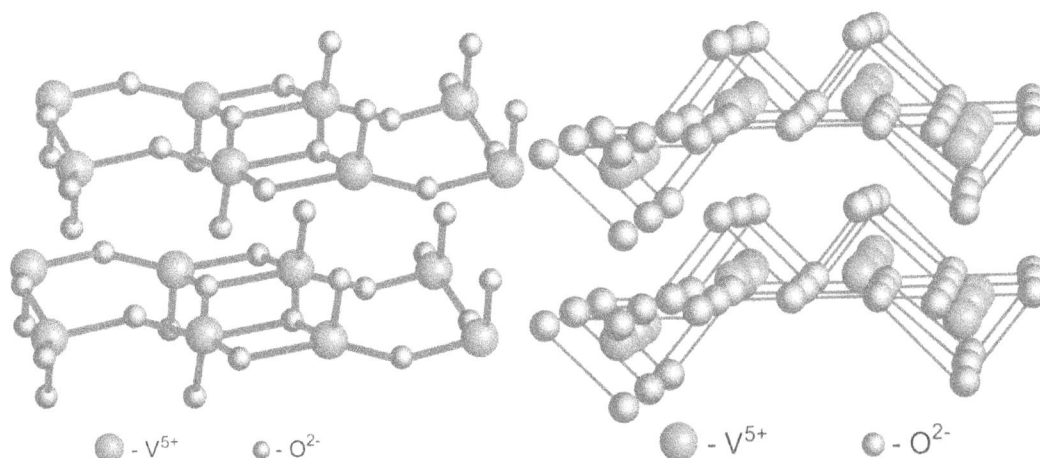

Fig.13 Perspective view of two layers of V_2O_5.

Its layered structure makes V_2O_5 a promising material for energy storage systems, and its high ionic storage capacity makes it a suitable dielectric constituent material in super capacitors [51]; it can also be applied in electrochromic [55] as well as optical switching devices [53], as a reversible cathode material for Li batteries [57-59], and as a thermo-resistive material in thermal infrared detectors [60].

V_2O_5 is a wide-band-gap, n-type semiconductor material [61] that exhibits particularly useful properties stemming from the fact that in this case charge is transported via polarons, as reported for both bulk single-crystal structure [62] and amorphous V_2O_5 layers [63, 64]. The surface properties of V_2O_5 as part of both ceramic sinters and thin films make it candidate for many practical applications, including catalyst materials and gas sensors [65-74]. The most typical application of V_2O_5 as a heterogenic catalyst is the fabrication of sulphuric acid, an important industrial chemical with an annual worldwide production of about 200 million tons. Its main purpose in this regard is to catalyze the oxidation of sulphur dioxide to sulphur trioxide by air [68].

V_2O_5 can be prepared via the direct oxidation of vanadium metal, but this product is contaminated with other lower oxides. The decomposition of ammonium metavanadate (NH_4VO_3) at around 470 K is a far more efficient preparation method. The methods used to prepare V_2O_5 thin films are described in chapter 4 of this issue. The oxide has a melting point of 963 K. Upon heating, it reversibly loses oxygen. The extent of non-stoichiometry x in V_2O_{5-x} varies between $0 \leq x \leq 0.01$ [69].

Vanadium pentoxide is a saturated oxide, i.e. vanadium ions have the highest oxidation state in this compound, and therefore it is the most stable one in the vanadium-oxygen system. It is known to be an excellent catalyst [6] owing to its rich and diverse chemical properties stemming from two factors: the possibility to change vanadium oxidation states, ranging from +2 to +5 at the surface, and the variability of oxygen coordination geometries. This structural richness is the source of the existence of oxygen ions with different coordination, which are a significant component in controlling physical and chemical surface properties [7]. V_2O_5-based catalysts are used to fabricate important chemicals and in environment-friendly technologies that reduce pollution [7]. Of all heterogeneous catalysts, V_2O_5 is the most frequently applied one (28%). Its unique surface properties make it a prospective material for the development of chemical sensors. Moreover, vanadium pentoxide thin films are used as reversible cathode materials in secondary Li batteries, electrochromic materials, bolometers etc. Because of its layered structure and weak bonds between the layers, it is possible to extract layers with a thickness at the nm level or even a monolayer of V_2O_5. Such a layer retains the same properties (e.g. metal-insulator transition) that are characteristic of a thin film of V_2O_5. Since certain properties of graphene (a single layer of graphite) are unique compared to those of its bulk counterpart, it is worth establishing whether such notable properties can be observed in the case of a single layer of vanadium pentoxide [8].

Vanadium oxides are also used in many technological applications, including electrical and optical switching devices, light detectors, thermal and chemical sensors, write–erase media, and in heterogeneous catalysis [9,10]. Studies on the solid-state physics of vanadium oxides concentrate on phase transitions, in particular metal-insulator transitions (MIT) as a function of temperature. MITs are associated with peculiar structural, electronic, and magnetic behaviour and certain questions concerning their theoretical description remain unanswered [11-19].

Layered V_2O_5 is one of the most attractive cathode materials in lithium-ion batteries (LIBs) [70]. It is characterized by a high theoretical capacity of 437 mAhg^{-1}, which is much higher than that of the currently used $LiCoO_2$ (227 mAhg^{-1}). In a Li/V_2O_5 half-cell LIB, the Li$^+$ intercalation and de-intercalation process can be expressed as follows:

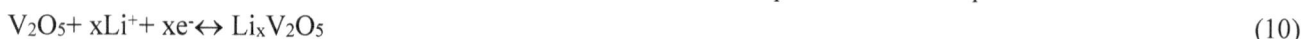

$$V_2O_5 + xLi^+ + xe^- \leftrightarrow Li_xV_2O_5 \tag{10}$$

5. MAGNÉLI AND WADSLEY HOMOLOGOUS VANADIUM OXIDES

Apart from the single-valence (principal) oxides, the phase diagram also includes mixed-valence oxides containing vanadium in two oxidation states. The mixed-valence oxides form due to changes in the deviation from stoichiometry in the principal oxides V_2O_3 and VO. When the concentration of the oxygen vacancy defects that are formed exceeds a certain threshold, the vacancies interact with each other, forming crystallographic shear planes; in other words, the vacancies associate along a lattice plane and are subsequently eliminated via the reorganisation of V-O coordination units [71]. Because of these interactions, two homologous series exist in the vanadium-oxygen system. Such series with a composition described with the general formula $V_nO_{2n-1} = VO_{2-1/n}$ are known as Magnéli phases, whereas Wadsley phases have the general formula $V_nO_{2n+1} = VO_{2+1/n}$.

Magnéli phases of vanadium oxides (V_nO_{2n-1}) can be considered chemical compounds in the system of principal oxides V_2O_3-VO_2, formed according to the reaction:

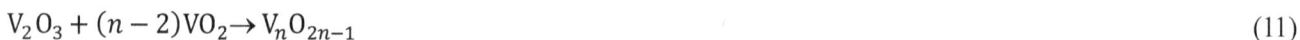

$$V_2O_3 + (n-2)VO_2 \rightarrow V_nO_{2n-1} \tag{11}$$

where n is an integer between $3 \leq n \leq 9$. For the two reagents (V_2O_3 and VO_2), the parameter n corresponds to values of 2 and $+\infty$, respectively.

As indicated by Eq. (11), Magnéli phases should exhibit properties that are intermediate between those of V_2O_3 and VO_2. Indeed, their crystal structure consists of rutile (VO_2) and corundum (V_2O_3) blocks. The structure of V_nO_{2n-1} includes nVO_2 units between shear planes. Consequently, V_nO_{2n-1} exhibits structural affinity to both VO_2 and V_2O_3. The detailed experimental studies of the structural and electronic properties of Magnéli phases performed by Schwingenschlögl and Eyert [73] confirmed this hypothesis.

In terms of structure, Magnéli phases are considered rutile-type (VO_2) 2D in finite slabs with a thickness of n VO_6 octahedra [74,75].

The chemical composition of Magnéli phases corresponds to a crystal structure composed of V^{3+} ([Ar]3d^2) and V^{4+} ([Ar]3d^1) cations and O (1s^{2}2s^{2}2p^2) anions. The chemical compositions of experimentally identified Magnéli phases are listed in Table 3 together with some of their other properties.

The strong correlation between the chemical composition and charge ordering (coordination numbers of the ions) in the crystal structure of the Magnéli phases had already been proved [71]. However, much less is known thus far about the Wadsley series of vanadium oxides. The chemical formula of the Wadsley vanadium oxide series is widely accepted to be V_nO_{2n+1} [71]. There are several reports which claim that it can be expressed as $V_{2n}O_{5n-2}$ [71-78]. However, the experimentally identified phases of this type, shown in Fig.1 – V_3O_7, V_4O_9, V_5O_{11}, V_6O_{13}, V_7O_{15} – indicate that V_nO_{2n+1} is the correct general formula. The Wadsley series can be considered the chemical product of a reaction between V_2O_5 and VO_2:

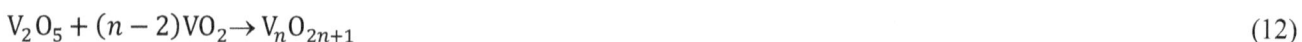

$$V_2O_5 + (n-2)VO_2 \rightarrow V_nO_{2n+1} \tag{12}$$

where n is an integer. For the experimentally identified Wadsley phases, $2 \leq n \leq 7$. For the two reagents (V_2O_5 and VO_2), the parameter n corresponds to value of 2 and $+\infty$, respectively.

Like vanadium pentoxide, Wadsley phases have a layered structure. These compounds permit reversible ion intercalation, which makes them suitable for electrochemical energy conversion and storage [79-86].

Table 3. Basic properties of the Magnéli oxides of vanadium (V_nO_{2n-1})

Material Formula	Para-meter n	Formal vanadium valence	Number of ions in a molecule			T_{MIT} [K]	Thermodynamic data at T_{MIT}		Crystal structure at RT	E_g [eV]
			V^{3+}	V^{4+}	O^{2-}		ΔH_{MIT} [J/mol] [87]	ΔS_{MIT} [J/(mol·K)] [87]		
V_2O_3	2	3.00	2	0	3	155 [88]	1490	9.01	D^6_{3d}-R3̄c, $Z = 6$, $a = 0.49517$, $c = 1.4005$ [89]	
V_3O_5	3	3.33	2	1	5	430 [87] 427 [90]	900	2.09	C^4_{2h}-P2/c [87] RT, $a = 1.0004$, $b = 0.5040$, $c = 0.9854$ nm, $\beta = 37.9°$ C^5_{2h}-P2$_1$/c [88] $a = 0.985$, $b = 0.50416$, $c = 0.6991$ nm, $\beta = 109.478°$	0.1
V_4O_7	4	3.50	2	2	7	250 [93] 240 [94] 238 [95]	595	2.39	C^1_i-A1̄ $Z = 4$ [94, 96, 97], $a = 0.5509$, $b = 0.7008$, $c = 1.2256$ nm, $\alpha = 95.10°$, $\beta = 95.17°$ $\gamma = 109.25°$	0.15
V_5O_9	5	3.60	2	3	9	135 [102, 95]	900	5.66	C1i-A1̄ $Z = 2$ [98], $a = 0.5472$, $b = 0.7003$, $c = 0.8727$ nm, $\alpha = 97.49°$, $\beta = 112.40°$, $\gamma = 109.01°$	0.2
V_6O_{11}	6	3.67	2	4	11	170 [93, 95]	929	5.48	C^1_i-P1̄ $Z = 2$ [92], $a = 0.5448$, $b = 0.6998$, $c = 3.0063$ nm, $\alpha = 41°$, $\beta = 72.5°$, $\gamma = 108.9°$	
V_7O_{13}	7	3.71	2	5	13	no MIT	-	-	C1i-P1̄ $Z = 2$ [92], $a = 0.5439$, $b = 0.7005$, $c = 3.5516$ nm, $\alpha = 40.9°$, $\beta = 72.6°$, $\gamma = 109.0°$	
V_8O_{15}	8	3.75	2	6	15	68 [95]			C1i-P1̄ $Z = 2$ [99], $a = 0.542$, $b = 0.702$, $c = 4.065$ nm, $\alpha = 40.8°$, $\beta = 73.2°$, $\gamma = 109.5°$	
V_9O_{17}	9	3.78	2	7	17	79 [93, 100, 87]			C1i-P1̄ $Z = 2$ [101], $a = 0.5418$, $b = 0.7009$, $c = 4.5213$ nm, $\alpha = 39.3°$, $\beta = 74.5°$, $\gamma = 108.9°$	

VO₂	$\propto$	4.00	0	1	2	340 [102] 341 [103]	4237 [87] 17865 [103]	12.56 [87] 52.388 [103]	C^5_{2h}–P2₁/c Z = 4 [104]	0.6 [105]

Research on compounds belonging to the Wadsley series is an important prerequisite for understanding the catalytic mechanism of V_2O_5 in its redox reactions [106]. In addition, such materials have found application in oxide electronics – as a dielectric in capacitors [107]. The chemical composition of experimentally identified Magnéli phases V_nO_{2n-1} is listed in Table 3.

Table 4 lists the chemical compositions and selected properties of Wadsley phases of vanadium oxides.

Table 4. Basic properties of the Wadsley oxides of vanadium (V_nO_{2n+1})

Material formula	Para-meter n	Formal vanadium valence	Number of ions in a molecule			T_{MIT} [K]	Low T Structure ($T < T_{MIT}$)	Crystal structure at RT	E_g [eV]
			V^{5+}	V^{4+}	O^{2-}				
V_2O_5	2	5.00	2	0	5	530	D^{13}_{2h}-Pmm $Z = 2$, $a = 1.1510$, $b = 0.4369$, $c = 0.3563$ [108]		2.25 $E\|c$, 2.23 $E\|a$ [109] 2.19 $E\|b$ [110]
V_3O_7	3	4.67	2	1	7	5	Monoclinic C_{2h}^6-C2/c [1117] 4K $a = 2.191$, $b = 0.368$, $c = 1.829$nm, $\beta = 95.7°$	Monoclinic C_{2h}^6 -C2/c [111] $a = 2.1921$, $b = 0.3679$, $c = 1.8341$ nm, $\beta = 95.61°$	
V_4O_9	4	4.50	2	2	9			Orthorhombic D_{2h}^7-Pmna $Z = 4$ [111, 112] $a = 1.7926$, $b = 0.3631$, $c = 0.9396$ nm $Z = 8$ [113] $a = 0.8235$, $b = 1.032$, $c = 1.647$ Tetragonal $Z = 4$ [114] $a = b = 0.8215$, $c = 1.032$ nm	
V_6O_{13}	6	4.33	2	4	13	151	Monoclinic C_{2h}^5-2P₁/ₐ[115,116,117] $a = 1.1963$, $b = 0.3707$, $c = 1.0064$ nm, $\beta = 100.96°$	Monoclinic C_{2h}^3-C2/m $Z = 2$ [117,118] $a = 1.1921$, $b = 0.6811$ $c = 1.0147$ nm, $\beta = 100.88°$	

VO₂	∝	4.00	0	1	2	340	-	C^5_{2h}-P2$_1$/c $Z = 4$ [104]	0.60 [105]

6. CONCLUSIONS

This paper presents a broad overview of single-valence vanadium oxides and mixed-valence variants which form the Magnéli and Wadsley homologous series. Under certain ambient conditions (including temperature), phase transformations between these oxides can occur. Based on the available thermodynamic data, the specific conditions required to obtain particular oxides were determined.

References

1. H.A. Wriedt, *The O-V (Oxygen-Vanadium) system*, in: Bulletin of Alloy Phase Diagrams Vol. 8 No. 2 1987.
2. J. Stringer, *The vanadium –oxygen system – a review*. Less Common Metals 8 (1965) 1-14.
3. C.H. Griffith, H.K. Eastwood, *Influence of stoichiometry on metal-semiconductor transition vanadium dioxide*, J.Appl.Phys. 45 (1974) 2201
4. N. Bahlawane, D. Lenoble, *Vanadium oxide compounds: Structure, properties and growth from the gas phase*, Chemical Vapour Deposition 20 (2014) 299-31.
5. I. Barin, J. Knacke, O. Kubaschewski, Thermochemical properties of inorganic substances. Supplement, Spiriner-Verlag, Berlin-Heidelberg-New York 2013.
6. K. Schneider, K. Zakrzewska, Z. Tarnawski, K. Drogowska, N-T.H. Kim-Ngan, *VOx thin films deposited by reactive rf sputtering,* Ceramics v. 115 (2013) 305-314.
7. Handbook of Chemistry and Physics, 60 Edition, CRC Press (1980) p. B-140.
8. B.M. Vasyutinskiy, A.L. Donda, G.N. Kartmazov, Yu.M. Smirnov, V.A. Finkel, *Structure data of elements and intermetallic phases*, Izv. Akad. Nauk SSSR, Metally, *Rus. Metall. (English Transl.),* 6 (1967) 84.
9. D.A. Davidov, A.A. Rempel. *Lattice parameter, density, and defect system of VO_y,* Inorg. Mater. 45 (2009) 666-670.
10. Y. Ueda, K. Kosuge, S. Kachi, J.Solid State Chem., *Phase diagram and some physical properties of V_2O_{3+x} ($0 \leq x \leq 0.080$)* J. Solid State Chem.31 (1980) 17188.
11. R.E. Word, S.A. Werner, W.B. Yelon, J.M. Honig, S. Shivashankar, *Spin waves in vanadium sesquioxide,* Phys.Rev. B 23 (1981) 3533.3540.
12. R. Belkbeoch, R. Kleinberger, M. Roulliay, *Lattice parameter anomalies in V_2O_3 at high temperature,* Solid State Comm. 25 (1978) 1043-1044.
13. A. Zylbersztejn, N.F. Mott (1975), *Metal-insulator transition in vanadium dioxide,* Phys Rev B 11 (1975) 4383-4395.
14. D. Kucharczyk, T. Niklewski, *Accurate X-ray determination of the lattice parameters and the thermal expansion coefficients of VO_2 near the transition temperature,* J. Appl. Cryst. (1979). 12, 370-373.
15. J.C. Parker, U.W. Geiser, D.J. Lam, Xu, W.Y. Ching, *Optical properties of vanadium pentoxide determined from elipsometry and band-structure calculations,* Phys.Rev. B 42 (1990) 5289-5293.
16. G. S. Nadkarmi, V.S. Shirodkar, *Experiment and theory for switching in $Al/V_2O_5/Al$ devices,* Thin Solid Films 105 (1983) 115-129.
17. H.G. Bauchmann, F.R. Ahmed, W.H. Barnes, *The crystal structure of vanadium pentoxide-SAO/NASA ADS* Z. Kristallogr.115(1961) 110-13.
18. V.G. Mokerov, *V_2O_5 property: energy gap E_g 2.17 eV,* Fiz.Tverd.Tela 15(1973) 2393.

19. D.S. Volzhensky, V.A. Grin, V.G. Savitskii, Kristallografiya 21(1976)1238.

20. J. Stringer, *The vanadium-oxygen system*, J. Less Common Metals 8 (1965) 14.

21. E.M. Page, S.A. Wass, Vanadium: Inorganic and Coordination chemistry, Encyclopaedia of Inorganic Chemistry, John Wiley &Sons, ISBN 0-4793620-0 (1994).

22. D.B. McWhan, T.M. Rice, *Critical pressure for the metal-semiconductor transition in V_2O_3,* Phys. Rev.Lett. 22 (1969) 887.

23. D.B. McWhan, T.M. Rice, J.P. Remeika, *Mott transition in Cr-doped V_2O_3*, Phys. Rev.Lett.23 (1969) 1384.

24. M. Wakihara, T. Katsura, *Thermodynamic properties of the V_2O_3-V_4O_7 system at temperatures from 1400° to 1700° K*, J.Phys.Chem. Soc. Japan 1 (1970) 363-366.

25. N.N. Greenwood, A. Earnshaw, *Chemistry of the Elements 2nd ed.* Butterworth-Heinemann, ISBN 0-08-037949 (1997).

26. C. Lamsal, N. M. Ravindra, *Optical properties of vanadium oxides- an analysis,* J. Mater. Sci. (2013) 48, 6341-6351.

27. P.Kofstad, *Nonstoichiometry, Diffusion and Electrical Conductivity in Binary Metal Oxides,* Wiley, Intersci., NY, London, Sydney, Tokyo, 197

28. Y.Y. Luo, F.H. Su, C. Zhang, L. Zhong, S.S. Pan, S.C. Xu, H. Wang, J.M. Dai, G.H. Lia,, *Terahertz transport dynamics in the metal-insulator transition of V2O3 thin film, Optics Commun.* 387 (2027) 385-389

29. Y. Guo and J. Robertson, *Analysis of metal insulator transition in VO_2 and V_2O_3 for RRAMs*, Microelectron. Eng. 109 (2013) 278-281.

30. S. Long, X. Cao, G. Sun, N. Li, T. Chang, Z. Shao, O. Jin, *Effects of V_2O_3 buffer layers on sputtered VO_2 smart windows: Improved thermochromic properties, tuneable width of hysteresis loops and enhanced durability,* Appl.Surf.Sci. 441 (2018) 764-772.

31. N. A. Charipar, H. Kim, S. A. Mathews, and A. Pique, *Formation energies of intrinsic point defects in monoclinic VO_2 studied by first-principle calculations,* AIP Adv. 6, 015113 (2016)9.

32. B. Rajeswaran and A. M. Umarji, *Effect of W addition on the electrical switching of VO_2 thin films,* AIP Adv. 6, 035215 (2016) 8.

33. M. M. Yang, Y. J. Yang, B. Hong, H. L. Huang, S. X. Hu, Y. Q. Dong, H. B. Wang, H. He, J. Y. Zhao, X. G. Liu, Z. L. Luo,X. G. Li, H. B. Zhang, and C. Gao, AIP Adv. 5, 037114 (2015). 105309.

34. A.V. Nikolaev, Yu.N. Kostrubov, B.V. Andreev, *The method of linear augmented plane waves (LAPW)*, Sov. Phys. Solid State 34 (1992) 1614-1620.

35. A. Stivastava,H. Rotella, S. Saha, B. Pal, G. Kalon, S. Methew, M. Matapothula, M. Dykas, P. Yang, E. Okunishi, D.D. Sarma, T. Venkatesan, *Selective growth of single phase (A, B and M) polymorph thin films,* APL Mater. 3 (2015) 02610.

36. D. Hagman, J. Zubieta, Ch. J. Warren, L.M. Meyer, M.J. Tracy, R.C. Haushalter, *A new polimprph of VO_2 prepared by soft chemical methods,* J.Solid State Chem. 138 (1998) 178-182.

37. S. Lee, I.N. Ivanov, J.K. Keum, H.N. Lee, *Epitaxial stabilization and phase instability of VO_2 polymorphs,* Sci. Rep. 6 (2015) 19621.

38. S.R. Popuri, A. Artemenko, C. Labrugere, M. Miclau, M. Pollet, *VO2 (A): Reinvestigation of crystal structure, phase transition and crystal growth mechanism,* J.Solid State Chem. 213 (2014) 79-86.

39. F. Theobald, R. Cabala, J. Bernard., *Essai sur la structure de VO2 (B),* J. Solid State Chem., 17 (1976) 43438.

40. G. Andersson, *Studies on vanadium oxides. II.The crystal structure of vanadium dioxide,* Acta Chem. Scand., 10 (1956) 623-628.

41. S. Westman, *Note on phase transition in VO_2*, Acta Chem Scand. 15 (1961) 217-217.

42. Y. Oka, S. Sato, T. Yao, N.J. Yamamoto, *Crystal structures and transition mechanism of VO2 (A)*, Solid State Chem., 141 (1998) 594-598.

43. F. H. Chen, L. L. Fan, S. Chen, G. M. Liao, Y. L. Chen, P. Wu, L. Song, C. W. Zou, and Z. Y. Wu, *Control of metal-insulator transition in VO2 epitaxial film by modifying carrier density,* ACS Appl. Mater. Inter. 7 (2015) 6875-6880.

44. H. Kim, N.S. Bingham, N.A. Charipar, A. Pique, *Strain effect in epitaxial VO2 thin films grown on sapphire substrates using SnO2 buffer layers,* AIP Advances 7. 105116 (2017) 10.

45. J. Jeong, N. Aetukuri, T. Graf, T. D. Schladt, M. G. Samant, and S. S. P. Parkin, Science 339 1402 (2013).

46. Y. Sun, S. Jiang, W. Bi, R. Long, X. Tan, C. Wu, S. Wei, and Y. Xie, *New aspects of size-dependent metal-insulator transition in synthetic single-domain monoclinic vanadium dioxide nanocrystals,* Nanoscale 3 (2011) 4394.440.

47. Q. Yu, W. Li, J. Liang, Z. Duan, Z. Hu, J. Liu, H. Chen, J. Chu, *Oxygen pressure manipulations on the metal–insulator transition characteristics of highly (0 1 1)-oriented vanadium dioxide films grown by magnetron sputtering,* J. Phys. D Appl. Phys. 46, 055310 (2013) 10.

48. Y. Zhao, C. Karaoglan-Bebek, X. Pan, M. Holtz, A.A. Bernussi, Z. Fan, *Hydrogen- doping stabilized metallic VO2 (R) thin films and their applications to supress Fabry-Perot resonances in the terahertz regime,* Appl. Phys. Lett 104 (2014) 24190241905.

49. H.T. Yuan-Tao, K.C. Feng, X.J. Wang, C. Li, C.J. He, and Y.X. Nie, *Effect of nonstoichiometry on Raman scattering of VO2 films,* Chin. Phys. 13 (2004) 887.

50. Y. Cui, B. Liu, L. Chen, H. Luo, Y. Gao, *Formation energies of intrinsic point defects in monoclinic VO2 studied by first-principles calculations,* AIP Advances 6 (2016) 15301.

51. S. Fan, L. Fan, Q. Li, J. Liu, B. Ye, *The identification of defect structures for oxygen pressure dependent VO2 crystal films,* App. Surf.Sci. 321 (2014) 464-468.

52. H.M.R. Giannetta, C. Calaza, D.G. Lamas, L. Fonseca, L. Fraigi, *Electrical transport properties of V2O5 thin films obtained by thermal annealing of layers grown by RF magnetron sputtering at room tempera*ture, Thin Solid Films 589 (2015) 730-734.

53. J.B. Goodenough, *Metallic Oxides,* Prog. in Solid State Chem. 5 (1971) 145-399.

54. P. Pasierb, M. Rekas., High-*temperature electrochemical hydrogen pumps and separators,* Int. J.Electrochem. 2011 (2011) 10.

55. J.G. Zhang, J.M. McGraw, J. Turner, D. Ginley, *Charging capacity and cycling stability of VOx ilms prepared by pulsed laser deposition,* J. Electrochem. Soc. 144 (1997) 1630-1634

56. G. Guzman, B. Yebka, J. Livage, C. Julien, *Lithium intercalation studies in hydrated molybdenum oxides,* Solid State Ionics 86-88 (1996) 407-413.

57. C. Julien, E. Haro-Poniatowski, M.A. Camacho-López, L. Escobar-Alarcón, J. Jímenez-Jarquín, *Lithium batteries: Science and Technology,* J. Mater. Sci. Eng.B 65 (1999) 170-176.

58. A. Kumar, P. Singh, N. Kulkarni, D. Kaur, *Structural and optical studies of nanocrystalline V2O5 thin films,* Thin Solid Films 516 (2008) 912-918.

59. A.-M. Cao, J.-S. Hu, H.-P. Liang, L.-J. Wan, *Self-assembled vanadium pentoxide (V2O5) hollow microspheres from nanorods and their application in lithium-ion batteries,* Angew. Chem. Int. Ed. 44 (28) (2005) 4391–4395.

60. P.W. Kruse, *Uncooled Thermal Imaging: Arrays, Systems, and Applications,* SPIE Press, Bellingham, Wash., USA, 200.

61. V. Eyert, K.-H. Hock, *Electronic structure of V2O5: role of octahedral deformations,* Phys. Rev. B 57 (20) (1998) 12727-12737.

62. V. Ioffe, I. Patrina, *Comparison of the small-polaron theory with the experimental data of current transport in V2O5,* Phys. Status Solidi 40 (1) (1970) 389-395.

63. C. Sanchez, R. Morineau, J. Livage, *Electrical conductivity of amorphous V2O5,* Phys.Status Solidi A 76 (2) (1983) 661-666.

64. J. Bullot, P. Cordier, O. Gallais, M. Gauthier, J. Livage, *Experimental determination of the disorder energy in amorphous V_2O_5 layers deposited from gels,* Phys. Status Solidi A 68 (2) (1981) 357-36.

65. M.P. Suryawanshi, J.G. King, A.V. Moholkar, *Fast response of sprayed vanadium pentoxide (V_2O_5) nanorods towards nitrogen dioxide (NO_2) gas detection*, Appl. Surf. Sci. 403 (2017) 540-550.

66. S.D. Han, H.G. Moon, M.-S. Noh, J.J. Pyeon, Y.-S. Shim, S.N. Ahm, J.-S. Kim, K.S. Yoo, Ch.-Y. Kang, *Self-doped nanocolumnar vanadium oxides thin films for highly selective NO_2 gas sensing at low temperature*, Sensors and Actuators B 241, (2017) 40-47.

67. K. Schneider, M. Lubecka, A. Czapla. *V_2O_5 thin films for gas sensor applications*, Sensors and Actuators B: Chemical. Vol. 236 (2016), 970-977.

68. J.P. Dunn, P.R. Koppula, H.G. Stenger, I.E. Wachs, *Oxidation of sulfur dioxide to sulfur trioxide over supported vanadia catalysts,* Appl. Catalysis19 (1998) 105-117.

69. O. Kubaschewski, H. Hopkins, *Oxidation of metals and alloys,* Butterworth, London 196

70. X. Liu, J. Zeng, H. Yang, K. Zhou, D. Pan, *V_2O_5-based nanomaterials: synthesis and their applications*, RSC Adv. 8 (2018) 4014-4031

71. S. Mrowec, in Defects and diffusion in solids, Elsevier Amsterdam-Oxford-New York 1980.

72. U. Schwingenschlögl and V. Eyert, *The vanadium Magneli phases*, Condensed Materials (2004).

73. U. Schwingenschlogl, V. Eyert, *The vanadium Magneli phases V_nO_{2n-1}*, Annal. Phys. 2004, 13, 475-510.

74. J. S. Anderson, B. G. Hyde, *Introduction to chemical and structural defects in crystalline solid* J. Phys. Chem. Solids 29 (2012) 283-333.

75. S. Yamazaki, Ch. Li, K. Ohoyama, M. Nishi, M. Ichihara, H. Ueda, Y. Ueda, *Synthesis, structure and magnetic properties of V_4O_9 –A missing link in binary vanadium oxides*, J. Solid State Chem. 183 (2010) 1496-1503.

76. A.D. Wadsley, *Nonstoichiometric metal oxides. Order and disorder, Nonstoichiometric Compounds* vol. 39 ch. 2 pp 23-36 (1963).

77. C. Lamsal, N.M. Ravindra, *Optical properties of vanadium oxides-an analysis*, J.Mater. Sci. 48 (2013) 634-635.

78. S. Beke, *A review of the growth V_2O_5 films from 1885 to 2010*, Thin Solid Films 519 (2011) 1761-177.

79. E. Baudrin, G. Sudant, D. Larcher, B. Dunn, J. M. Tarascon, *Preparation of nanostructured VO_2 [B] from vanadium oxide aerogels,* Chem. Mater. 18 (2006) 4369-4374.

80. M.E. Arroyo-de Dompablo, U. Amador, J.M. Gallardo-Amores, C. Baehtz, N. Biskup, E. Morán, *High pressure materials for energy storage,* J.Phys:Conf.Series; (2008) V121, 032001.

81. Y. Wang, K. Takahashi, K. Lee, G.Z. Cao, *Nanostructured vanadium oxide electrodes for enhanced lithium-ion intercalation,* Adv.Funct.Mater. 16 (2006) 1133-1144.

82. N. A. Chernova, M. Roppolo, A.C. Dillon, M.S. Whittingham, *Layered vanadium and molybdenum oxides: batteries,* J. Mater. Chem. 19 (2009) 2526.

83. N. Peys, Y. Ling, D. Dewulf, S. Gielis, C. De Dobbelaere, D. Cuypers, P. Adriaensens, S. Van Doorslaer, S. De Gendt, A. Hardy, *V_6O_{13} films by control of the oxidation state from aqueous precursor to crystalline phase,* Dalton Trans.42 (2013), 959-968.

84. H. Li, P. He, Y. Wang, E. Hosono, H. Zhou, *Nano active materials for lithium-ion batteries,* Nanoscale 2 (2010) 1294-1305.

85. Y. Muranushi, T. Miura, T. Kishi, T. Nagai, *Insertion of lithium into vanadium molybdenum oxides,* Mater. Sci.; J.Power Sources (1987); Denki Kagaku 1986, 54, 691 (in Japanese).

86. A. Hammouche, A. Hammou, *Lithium insertion into V_4O_9* Electrochim. Acta 1987, 32, 1451.

87. F.A. Chodnovskii, and E.I. Tornukov, *Insulator-metal transition in V_3O_5,* Solid State Commun. 25 (1978) 573-577.

88. S. Shin, Y. Tezuka, T. Kinofhita, T. Ishii, T. Kashiwakura, M. Takahashi and Y. Suda, *Resonant soft X-ray study of rutile (TiO_2)* J.Phys.Soc.Jpn. 64 (1995).

89. R. Belbeoch, R. Kleinberger, and M. Roulliay, *Lattice parameter in V_2O_3 at high temperature,* Solid State Commun. 25 (1978) 1043-1044.

90. S. Asbrink, and S.-H. Hong, *Increase of X-ray reflection intensities and profile widths at the low-to high V_3O_5 phase transition state,* Nature (London) 279 (1979) 624-625.

91. S. Asbrink, *Studies on the system V_3O_5 /1b Ti_3O_5,* Mat. Res. Bull, (1975).

92. H. Horiuchi, N. Morimato, and M. Tokonami, J. Solid State Chem. 17, 407 (1976).

93. S. Kachi, K. Kosuge, and H. Okinaka, J. Solid State *A new polymorph of VO_2 prepared by soft chemical methods, J. Solid State* Chem. 6 (1973).

94. M. Marezio, D.B. McWhan, P.D. Dernier, J.P. Remeika, *Structural aspects of the metal-insulator transitions in V_4O_7* J. Solid State Chem. 6 (1973) 213-221.

95. S. Nagata, P.H. Keesom, and S.P. Faile, *Susceptibilities of the vanadium Magneli phases V_nO_{2n-1} at low temperature,* Phys.Rev. B20 (1979) 2886.

96. L.-J. Hodeau, and M. Marezio, *Structural aspects of the metal-insulator transitions in $(Ti_{0.9975}V_{0.0025})_4O_7$,* J. Solid State Chem. 29 (1979) 47-62.

97. H. Horiuchi, M. Tokonami, N. Morimoto, N. Nagasawa, *The crystal structure of V_4O_7* Acta Crystallogr. B28 (1972) 1404 -1410.

98. M. Marezio, P.D. Dernier, D.B. Wham, and S. Kachi, *Structural aspects of the metal-insulator transition in V_5O_9* J.Solid State Chem. 11 (1974) 301.

99. H. Horiuchi, M. Tokonami, N. Nagasawa , N. Morimoto,, Y. Bando, and T. Takada, *Crystallography of V_nO_{2n-1} ($3 \leq n \leq 8$)* Mater.Res.Bull. 6 (1971) 833-843.

100. S. Nagata, P.H. Keesom, H. Kuwamoto, S. Otsuka, and H. Kato, *Magnetic susceptibility of V_9O_{17} between 1 and 120 K,* Phys.Rev. B23 (1981) 411.

101. H. Kuwamoto, N. Otsuka, and H. Sato, *High- resolution electron microscopy of microsyntactic,* J. Solid State Chem. 36 (1981) 133.

102. G. Anderson, Acta Chem. Scand. *Studies on vanadium oxides. II. The crystal structure of vanadium dioxde, Acta Chem. Scand.* 10 (1956) 623-628.

103. C. Blaauw, F. Leenhouts, F. van der Woude and G.A. Sawatzky, *VO_2: resistivity, conductivity, photoconductivity,* J.Phys. C8 (1975) 459.

104. I. Barin, J. Knacke, O. Kubaschewski, Thermochemical properties of inorganic substances. Supplement, Springer-Verlag, Berlin-Heidelberg-New York 2013.

105. A. Gavini, and C.C,Y. Kwan, *VO_2: optical properties, dielectric constants,* Phys. Rev. B5 (1972) 3138 - 3143.

106. S.-G. He, Y. Xie, S. Heinbuch, E. Jakubikova, J.J. Rocca, E.R. Bernstein, *Reactions of sulphur dioxide with natural vanadium oxide clusters in the gas phase, II Experimental study employing single-photon ionization,* J.Phys.Chem. A 112 (2008) 11067-11077.

107. *Capacitor with catalogue number V_7O_{15}* Catalogue Aveston Electronics (in Russian).

108. H.G. Bachmann, F.R. Ahmed and W.H. Barnes, *The crystal structure of vanadium pentoxide;* Z. Kristallogr. 5 (1961) 110.

109. V.G. Mokerov, *V_2O_5: energy gap,* Fiz.Tverd. Tela 15 (1975) 2393-2395.

110. D.S. Volzenski, V.A. Grim, and V.G. Savitskii, Kristallografiya 21(1976) 1238.

111. M. Bayard, J.C. Grenier, M. Pouchard, and P. Hagenmmuller, *$V(n)O(2n+1)$ ($n>=5$) physical properties,* Mater. Res. Bull. 9 (1974) 1137-1144.

112. K.A. Wilhelmi, and K. Waltersson, *On the structure of a new vanadium oxide, V_4O_9* Acta Chem. Scand. 24 (1970), 3409-3411.

113. G. Grymonprez, L. Fiermans, and J. Vennik, *Structural properties of vanadium oxides,* Acta Crystallogr. A33 (1977) 834-837.

114. F. Theobald, R. Cabala, and J. Bernard, *Essai sur la structure de VO$_2$(B)* C.R. Acad.Sci. C 269 (1976) 431-438 .

115. M. Saeki, N. Kimizuka, M. Ishii, I. Kawada, and M. Nakahira, *Phase transition pf V$_6$O$_{13}$,* J. Less Common Met. 32 (1973) 171-172.

116. N. Kiwizuka, M. Nahano-Onoda, and K. Kato, *Structural re-investigation of the low-temperature;* Acta Crystallogr. B34, (1978) 1037-1039.

117. M. Saeki, N. Kimizuka, M. Ishii, I. Kawada, A. Ichinose, and M. Nakahira, *Crystal growth of V$_6$O$_{13}$,* J.Cryst. Growth 18 (1973)101 -102.

118. P.D. Dernier, *Structural investigation of the metal-insulator transition in V$_6$O$_{13}$,* Mater.Res.Bull. 9 (1974) 955-963 \.

III. Metal-insulator transitions in vanadium oxides

Abstract

This article presents state-of-the-art metal-insulator transitions (MITs) in vanadium oxides. MITs in the principal oxides V_2O_3, VO_2 and V_2O_5 are addressed in particular detail because of their practical importance. The review describes theoretical approaches to the MIT phenomenon. The presented information on MITs in vanadium pentoxide is based on the author's own experimental work.

Key words: Metal-insulator transition, mechanism of MIT, V_2O_5 ceramics, V_2O_5 thin films, MIT applications

1. INTRODUCTION

One of the most spectacular phenomena that occur in vanadium oxides is the abrupt change in electrical conductivity – from one typical of semiconductors to that typical of metal phases. This is known as a semiconductor-metal phase transition (SMPT) [1] or – more frequently – as a metal-insulator transition (MIT). Single-valence V_2O_3 and VO_2 as well as all Magnéli and Wadsley phases undergo a metal-insulator transition, except for V_7O_{13}, which is metallic at all temperatures. There are numerous controversies concerning the occurrence of the MIT in vanadium pentoxide (V_2O_5). Pergament et al. [2] argued that the term metal-insulator transition in vanadium pentoxide is not entirely correct. On the other hand, Kang et al. [3] and Kamper et al. [4] estimated the temperature at which the transition is observed in this compound to be 530 ± 5 K. Blum et al. [5] reported that a vanadium pentoxide (001) single crystal surface shows a reversible MIT and that this transition is limited to the surface layer.

2. MECHANISM OF A METAL-INSULATOR TRANSITION (MIT)

Hartree-Fock band approximation in quantum chemistry based on non-interacting electron systems successfully explains the distinction between metals, semiconductors and insulators. However, this theory fails to explain the electrical properties of transition metal oxides. Many transition metal oxides are poor conductors and often act as insulators in spite of the fact that they have a partially filled d-band.

Mott first explained this fact in 1949 [6]. According to him, the phenomenon is due to the electron-electron correlation, since electrons cannot be seen as non-interacting. Due to the interactions, electrons undergo strong Coulomb repulsion, which Mott [6] argued splits the band in two. At low temperatures, the lower band is completely filled and the upper band completely empty and the system is called a Mott insulator. If the temperature is high, a sufficient number of electrons will have enough energy to escape their site and become conduction electrons, allowing the flow of current. The result is that the material is either insulating or conductive depending on whether the temperature is low or high, respectively. Mott stated that the transition must be sudden, occurring when the density of free electrons (N) and the Bohr radius (r_B) satisfy the rule $N^{1/3}r_B \approx 0.2$. A Mott transition is a change in a material's properties from insulating to metallic due to various factors [7, 8].

The quantum mechanics theory that would explain the effect of electron interaction is very complex, because it is a many-body problem, which in general cannot be solved. In 1963, Hubbard proposed a simple approximate method. His model is based on perturbation theory [9]. According to the Hubbard model, the Hamiltonian has two components. The first component is the hopping integral representing the kinetic energy of electrons hopping between ions. The second component is the potential energy arising from the charges on the electrons, represented by parameter U that expresses repulsion.

At lower temperatures, Coulomb repulsion leads to the localization of electrons. As a result, the band d splits into two sub-bands, the lower (LH) band that is completely filled and the upper (UH) one, which is empty. Both sub-bands are limited by the Hubbard charge gap – E_g (Fig. 1). The oxide becomes an insulator. An MIT occurs when the thermal energy ($k_B T$) exceeds the Coulomb repulsion energy.

Fig. 1 presents a scheme of the Mott-Hubbard model. As can be seen, the p-band remains unchanged. On the other hand, at lower temperatures the interaction U prevails. The strong Coulomb repulsion leads to the localization of electrons. This results that the band d splits into two sub-bands, lower (LH) completely filled and upper (UH) empty. Both sub-bands are limited by Hubbard charge gap, E_g. The oxide becomes an insulator. Transition MIT occurs when thermal energy $k_B T$ surpass Coulomb repulsion energy.

The correlated electrons in vanadium oxides are responsible for their extreme sensitivity to external stimuli (excitations). The approach employed most frequently to induce MIT is via thermal excitation (i.e. changing the temperature by heating or cooling). This and the remaining approaches to inducing MITs are listed in Table 1 [10-15]. Fig. 2 illustrates the changes in resistance as a result of MITs triggered by the afore-mentioned excitations.

The mechanism of metal-insulator transitions in V_2O_3 is considered to be consistent with a canonical Mott-Hubbard model [16]. However, in other vanadium oxides, the mechanism of MITs is still under debate [17].

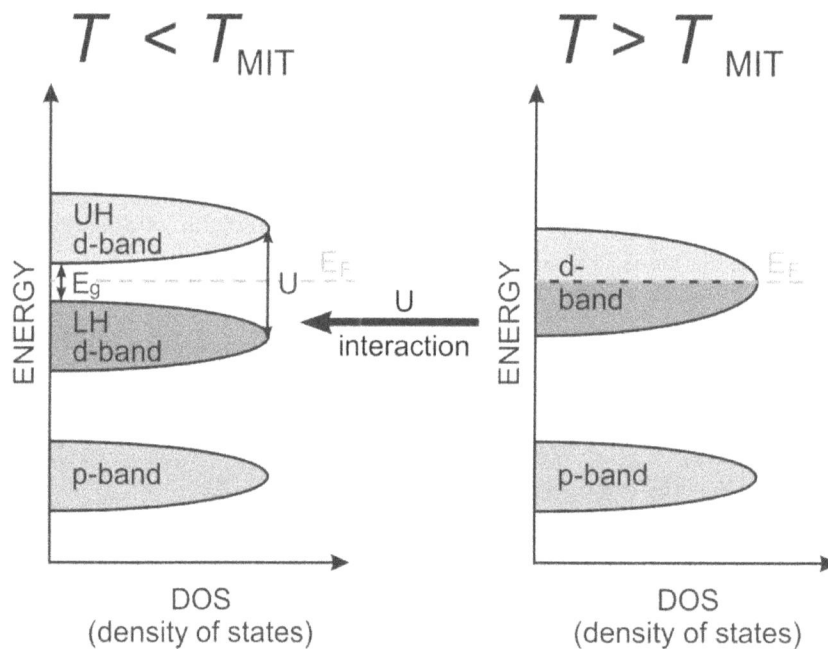

Fig. 1 Mott-Hubbard model of the MIT.

Another explanation of MITs was proposed Anderson [18]. He postulated that it is not electron interaction, but lattice disorder that leads to electron localization. The Anderson model is used to explain the electrical properties of heavily doped semiconductors [19].

According to recent opinions, both disorder and correlation play a role in real materials such as Na_xWO_3, Si:P. The main features of both Mott-Hubbard and Anderson MIT models are presented in Table 1.

Table 1 Features of MIT models

Mott-Hubbard model	Anderson model
electrical conductivity $\sigma = e^2 N(E_F)\mu$	
Mott insulator: $N(E_F) = 0$	Anderson insulator: $\mu = 0$
metal-insulator transition driven by:	
electron correlation (many-electron problem)	disorder (single-electron problem)
character of metal-insulator transition	
First-order transition tuned by external stimuli (excitations), mostly thermal (change in temperature), but also other, including: • electrical [10, 11] • optical [10, 12, 13] • magnetic [10] • strain [14] • chemical [15]	Continuous transition. Occurs at T close to 0 K and cannot be tuned by changing the temperature.

N – density of states (DoS), E_F – Fermi energy, μ – mobility of electron charges

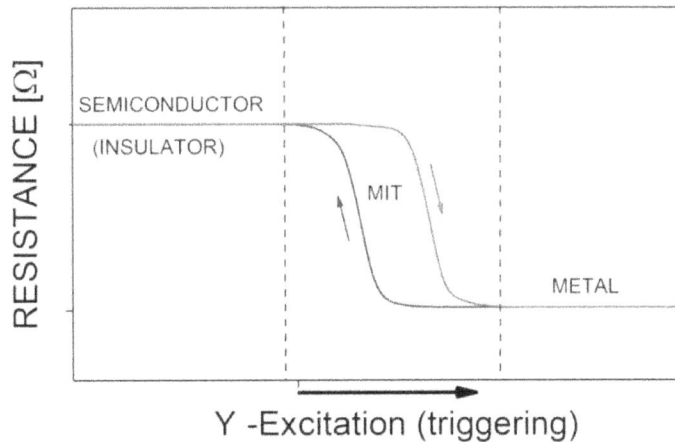

Excitation Y:
• Thermal, temperature [K]
• Electrical, applied voltage [V]
• Optical, laser pulse [photons/ps]
• Magnetic, magnetic field [T]
• Strain, stress [GPa]
• Chemical, gas concentration [ppm]

Fig. 2 Changes in resistance as a result of MITs triggered by external stimuli.

3. MIT IN VANADIUM OXIDES

MITs have been confirmed to occur in single-valence vanadium oxides VO_2, V_2O_3 and V_2O_5. Morin [20] and Austin [21] reported an MIT at about 123 K in VO, but other studies [22, 23, 24] have not confirmed this. MITs have also been observed for almost all Magnéli phases (with the single of exception of V_7O_{13}) and the V_6O_{13} Wadsley phase (data for other Wadsley phases). At the transition temperature (T_{MIT}), resistivity changes by as much as several orders of magnitude

(Fig. 3).

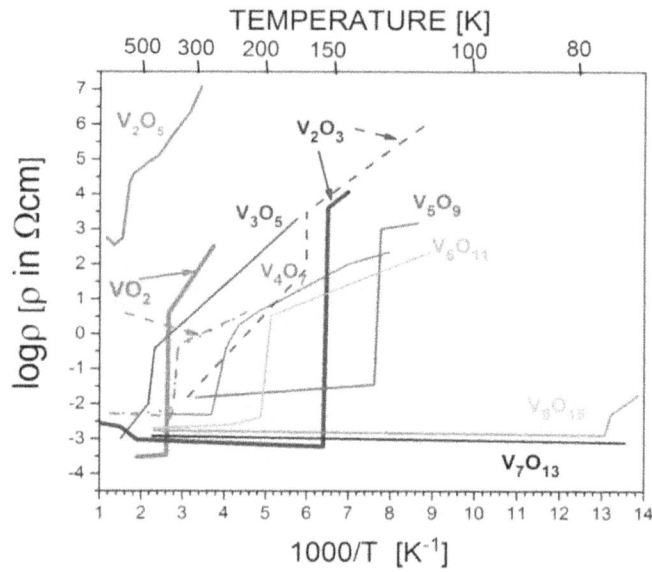

Fig. 3 Metal-insulator transition in vanadium oxides; sources of data for the graph: dashed lines [25, 26], line representing V_2O_5 [3], remaining lines [27].

Fig. 4 presents T_{MIT} as a function of x in VO_x based on the available data for vanadium oxides. No simple, approximate analytical formula for this dependence can be proposed.

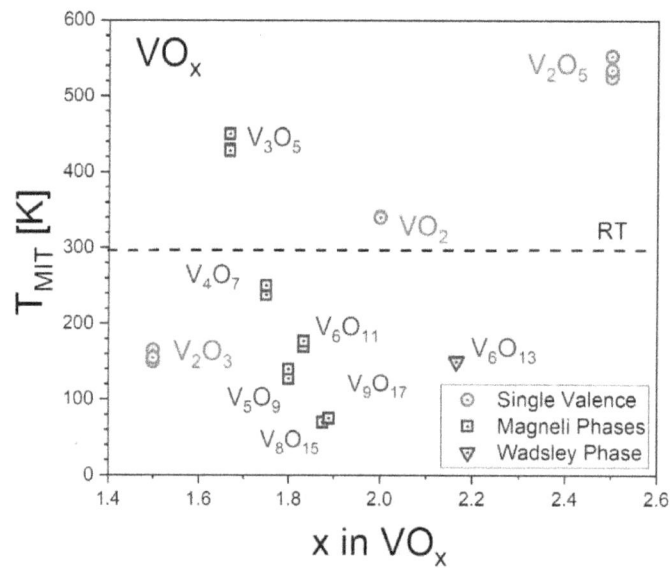

Fig. 4 Dependence of transition temperature (T_{MIT}) on the chemical composition of vanadium oxides (expressed as x in VO_x) [3, 25-28].

If single-valence vanadium oxides and Magnéli phases are considered separately (Figs 5 and 6), however, such formulae can be proposed.

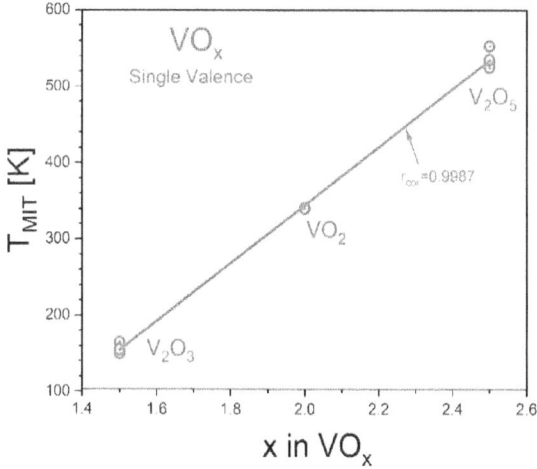

Fig. 5 T_{MIT} as a function of x in VO_x for single-valence vanadium oxides, plot after linear fitting.

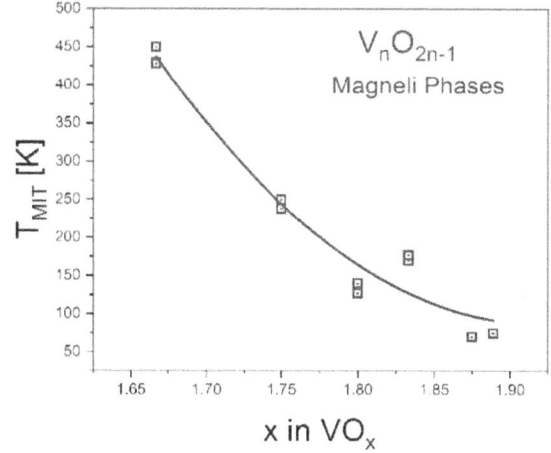

Fig. 6 T_{MIT} as a function of x in VO_x for Magnéli phases, plot after parabolic fitting.

The differences between the dependence T_{MIT} vs. x as seen in Figs 5 and 6 stem from the different physical meaning of x. In the case of single-valence oxides (Fig. 5), variable x determines the valence of V^{Z+} ions in V_2O_3, VO_2 and V_2O_5 ($Z = 3$, 4 and 5, respectively) and $x = Z/2$. On the other hand, Magnéli phases (V_nO_{2n-1}) can be treated as VO_2-V_2O_3 solid solutions composed of $(n - 2)$ moles of VO_2 and 1 mole of V_2O_3, with $x = 2 - \frac{1}{n}$.

Wedsley phases (V_nO_{2n-1}) can likewise be treated as VO_2-V_2O_5 solid solutions composed of $(n - 2)$ moles of VO_2 and 1 mole of V_2O_5, with $x = 2 + \frac{1}{n}$.

4. MIT IN V_2O_3

Vanadium sesquioxide (V_2O_3) is generally regarded as a canonical Mott-Hubbard system [16]. This oxide is among non-stoichiometric compounds that exhibit a deficit in the vanadium sublattice [29, 30]. Fig. 7 illustrates temeperature changes electrical resistance R and thermopower in $V_{2-x}O_3$ [31] and [32], respectively.

The T_{MIT} of stoichiometric V_2O_3 is (150.0 ± 0.5) K. A considerable effect of the vanadium deficit (x) on T_{MIT} is observed, with $T_{MIT} = (98.4 \pm 1.0)$ K and $T_{MIT} = (56.8 \pm 0.5)$ K for $x = 0.011$ and 0.019, respectively. The metal-insulator transition is accompanied by an immense change in resistivity – over 6 orders of magnitude. In addition, a drop in thermopower – from positive values at lower temperures ($T < T_{MIT}$) to negative ones at higher temperatures ($T > T_{MIT}$) – is observed.

The chemical composition of the sample used for thermoelectrical mesurements, as estimated from Fig. 7, is $V_{2-0.004}O_3$.

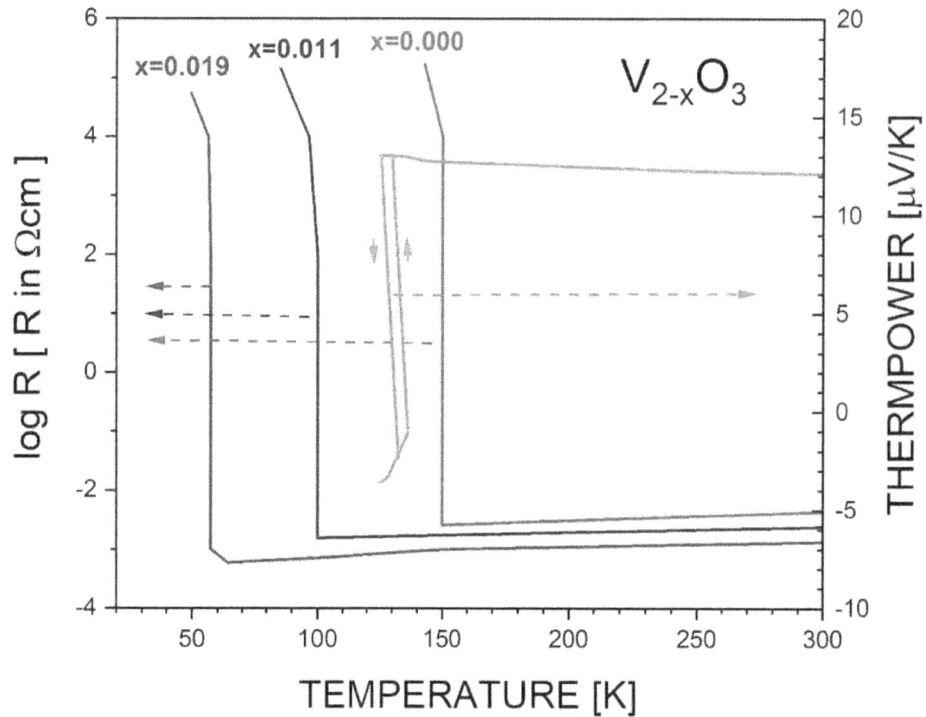

Fig. 7 Resistivity (R) and thermopower vs. temperature for $V_{2-x}O_3$ [30, 31].

5. MIT IN VO$_2$

Vanadium dioxide (VO$_2$) is an extensively studied material owing to its technological significance. In this case, MIT occurs at temperature close to room temperature. Other than via change in temperature, T_{MIT} can be tuned optically, electrically and via doping [32]. It is currently considered the most suitable material for electrical and optical switches [33] as well for oxide electronics such as the Mott-transition field-effect transistor, MTFET [34].

In the case of VO2, metal-insulator transition is usually accompanied by structural transformation and a subsequent change in the number of ions in the primitive unit cell.

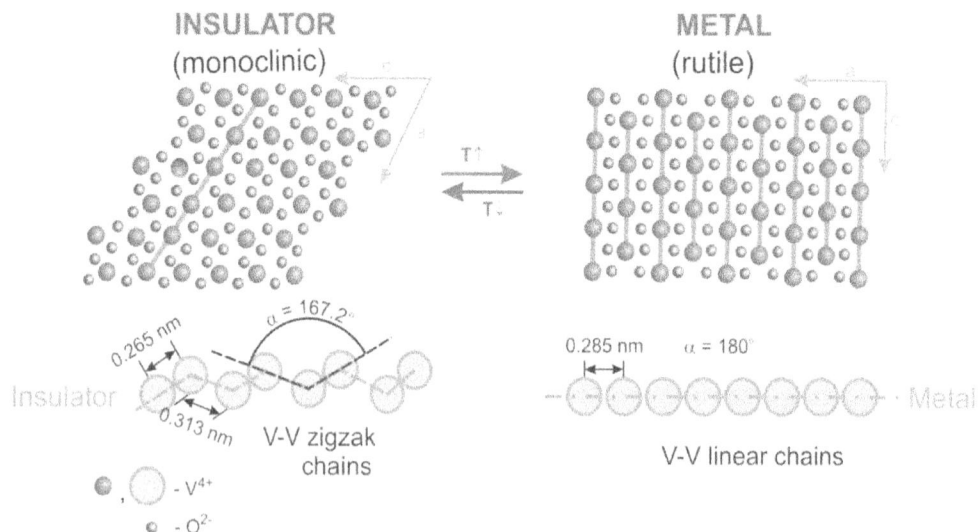

Fig.8 Projection of the [010] crystal plane of VO$_2$: left – an insulator state (monoclinic), right – a metal state (rutile).

Examples of the structures of the [010] crystal planes of VO_2 as an insulator (at temperatures below 340 K, i.e. below the T_{MIT}) and in a metallic state (i.e. above T_{MIT}) are shown in Fig. 8. In VO_2, the transition from an insulator to a metallic state is associated with a change from a monoclinic structure at low temperatures to the rutile structure of higher symmetry.

Since the discovery of the MIT in VO_2 [10], this kind of phase transition has been of major interest, as it is accompanied by a pronounced change in electrical resistivity and optical properties. This transition occurs at T_{MIT} = 340 K. For single crystals, the change in resistivity reaches a factor of 10^5 over a temperature range of 0.1 K [35]. The hysteresis associated with this transition is about 2 K. The conductivity jump and the narrowness of the hysteresis loop is a very good indication of how close the stoichiometry is to stoichiometric VO_2. Small deviations destroy the sharpness of the transition and increase the width of the hysteresis loop. The crystalline state of the material is significant, as well – a polycrystalline material will have a broader transition than single crystals. The transition temperature also depends on the crystalline state and oxygen non-stoichiometry. The results of experiments conducted by Morin [20] show the hysteresis of the resistivity of VO_2 to be about 26 K. Fig. 9 presents the temperature dependences of resistivity [5, 36] and optical transmittance reported for VO_2 thin films (thickness of 500 nm [36] and 1.55 Ωm [37]).

A)   B)

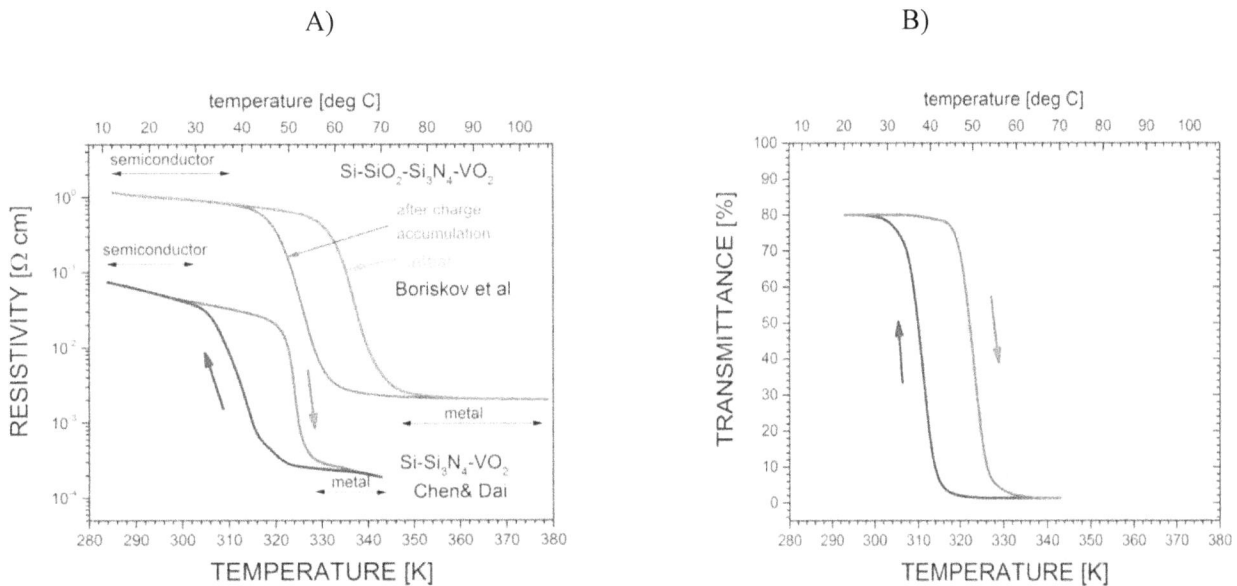

Fig. 9 Temperature dependence on: A – resistance [5,36] and B – optical transmittance [36], as reported for VO_2.

Chen and Dai [36] studied both electrical (Fig. 9A) and optical (Fig. 9B) switches made of VO_2 thin films with a thickness of 500 nm, deposited on a Si substrate with a 1 μm buffer layer consisting of Si_3N_4. The presented plots show hysteresis during heating and cooling; both electrical and optical characteristics indicate an MIT at 318 K (45°C).

Boriskov et al. [6] argued that the thermal effects observed for VO_2 during heating and cooling can be affected by the field effect in planar structures: $Si-SiO_2.Si_3N_4/ VO_2$ (Fig. 9A), the effect of electric field on the MIT is observed.

6. MIT IN V_2O_5

As in the case of VO_2, for other vanadium oxides the metal-to-insulator transition accompanies the phase transition from a lower to a higher symmetrical crystal structure. However, the vanadium pentoxide (V_2O_5) thin film is an exception. Vanadium pentoxide is a saturated oxide, which means that in this compound vanadium ions have the highest oxidation state. This makes V_2O_5 the most stable compound in the vanadium-oxygen system.

V_2O_5 exhibits highly anisotropic electrical and optical properties and it retains the same orthorhombic structure both below and above T_{MIT}.

Various transition temperatures in V_2O_5 have been reported in the literature. Wu et al. [37] studied the MIT in nanowires consisting of V_2O_5 doped with K, Cu, Na and/or W. They observed multiple jumps in resistance values and hysteresis between 300 and 400 K, suggesting that the MIT may occur in several steps from one phase to another with temperature changes. Moreover, the T_{MIT} was found to depend on the composition, length and diameter of the V_2O_5 nanowires. Kang et al. [3] and Blum et al. [5] estimated the T_{MIT} in V_2O_5 to be 530 ± 5 K. Blum et al. [5] reported that a vanadium pentoxide (001) single-crystal surface exhibits a reversible MIT and that this transition is limited to the surface layer.

According to Pergament et al. [27], the MIT in V_2O_5 is neither temperature-induced nor composition-induced as a result of the transition from V_2O_5 to nonstoichiometric V_2O_{5-x}. This transformation is instead associated with a reduction of vanadium pentoxide to lower oxides, such as V_6O_{13} or V_2O_3. Pergament et al. [2] argued that the term metal-insulator transition in vanadium pentoxide is not entirely correct. Furthermore, the precise mechanism of MIT in V_2O_5, if this transition does indeed occur, is still a matter of debate [38] and no theoretical understanding that would allow the transition temperature to be predicted has been achieved [39].

6.1 Study of MIT in V_2O_5 ceramics

An investigation of the electrical properties of polycrystalline vanadium pentoxide (V_2O_5) is the subject of this section. It is based on own studies, which are described in detail elsewhere [40]. X-ray diffraction (XRD) and scanning electron microscopy (SEM) were used to examine the structure and morphology of the investigated material. Electrical properties as a function of temperature and oxygen activity were determined by means of electrochemical impedance spectroscopy (EIS).

6.1.1 Experimental

Sample preparation

The reagent V_2O_5, produced by POCh (Gliwice, Poland), was used to prepare specimens for structural and electrical studies. The reagent was uniaxially pressed under a pressure of ca. 50 MPa and sintered in air at 673 K for 4 h. Prior to the impedance measurements, Ag paste was applied as the electrode on two surfaces of the obtained disc pellets (diameter ϕ = 10.01 mm, thickness d = 2.05 mm).

Structural and morphological characteristics

The phase composition of the samples was determined via XRD analysis conducted using a Philips X'Pert Pro diffractometer within the range of diffraction angles (2θ) from 20 to 100 deg., with monochromatic CuKα radiation λ = 0.154056 nm. Crystallite sizes were determined from the X-ray broadening of selected peaks, calculated by means of the Scherrer approach.

Microstructural observations and chemical analyses were carried out by means of scanning electron microscopy (SEM NOVA NANOSEM 200 FEI Europe Company) coupled with X-ray energy dispersive spectroscopy (EDAX company apparatus). For morphological observations, the samples were polished and subsequently thermally treated at 673 K for 1 h.

Impedance measurements

Electrical properties were determined by means of impedance spectroscopy, using a computer-controlled Solartron 1260 frequency response analyzer and a 1294 dielectric interface. The impedance spectra were analysed using the ZPLOT software package provided by Solartron. The measurements were performed within the frequency range between 0.1 Hz and 1 MHz. The amplitude of sinusoidal voltage was 10 mV.

6.1.2 Results and discussion

Structural and microstructural characteristics

Fig.10 presents the typical XRD patterns of the sample annealed in an argon atmosphere at several temperatures. X-ray diffraction analysis of the samples revealed the presence of the V_2O_5 orthorhombic phase.

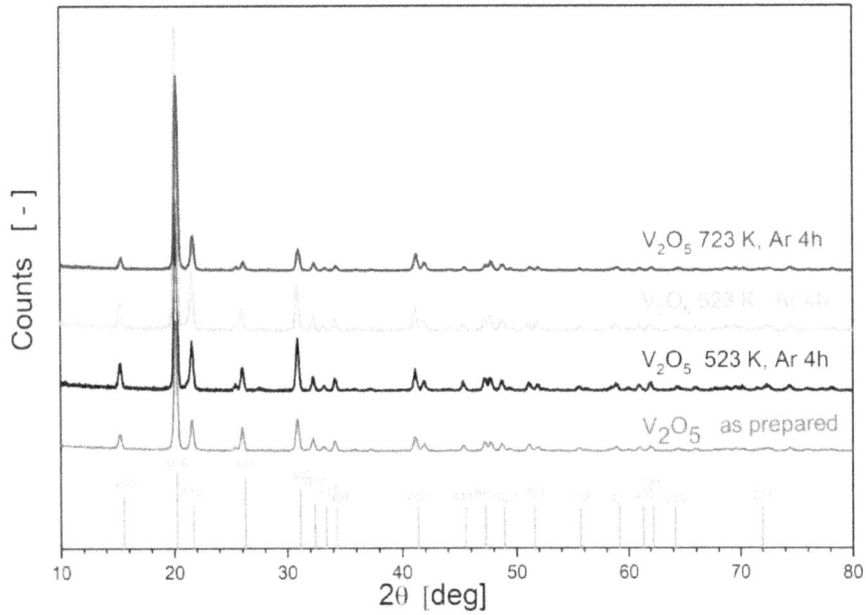

Fig. 10 X-ray diffraction patterns for V_2O_5 samples, as-prepared and sintered in argon at several temperatures.

Table 2 XRD analysis results of as-sputtered and annealed thin films

SAMPLE	Crystal structure	Lattice parameter			d_{XRD} [nm]
		a [nm]	b [nm]	c [nm]	
As-prepared	V_2O_5 Orthorhombic P_{nmm}	1.159 ± 0.002	0.438 ± 0.003	0.355 ± 0.008	0.466 ± 0.052
Ar, 523 K, 4h		1.158 ± 0.002	0.438 ± 0.002	0.355 ± 0.009	0.447 ± 0.034
Ar, 623 K, 4h		1.158 ± 0.001	0.438 ± 0.003	0.354 ± 0.007	0.490 ± 0.037
Ar, 723 K, 4h		1.152 ± 0.002	0.437 ± 0.004	0.352 ± 0.009	0.440 ± 0.073
Ref. [41]		1.148	0.436	0.355	-
Ref. [42]		1.1519	0.4373	0.3564	-
Ref. [43]		1.1512	0.4368	0.3564	-
10% H_2/Ar, 723 K, 4h	V_2O_3 Rhombo-	$a = b = c$ [nm]		α [rad]	0.315 ± 0.002
		0.546 ± 0.002		0.9477 ± 0.0009	

| Ref. [41] | hedral R-3c | 0.543 | 0.940 | - |

The determined lattice parameters are highly consistent with those reported in the literature [41, 42, 43]. The presented XRD patterns were used to determine the crystallite size (d_{XRD}), which was calculated according to Scherrer's method:

$$d_{XRD} = \frac{0.9\lambda}{\Delta(2\theta)*\cos\theta}$$ (1)

where $\lambda = 0.154056$ nm is the applied wavelength (CuK$_\alpha$), $\Delta(2\theta)$ denotes the broadening of the XRD peak at half of its maximum intensity, and θ represents the Bragg diffraction angle. The d_{XRD} parameter was determined for the eight peaks with the highest amplitude. The determined values of d_{XRD} are listed in Table 2. No effect of sintering temperature on the obtained XRD results was observed.

A
B

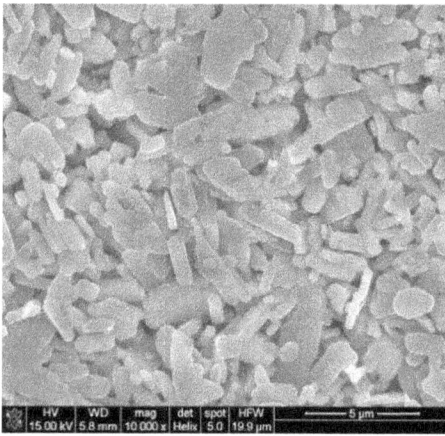 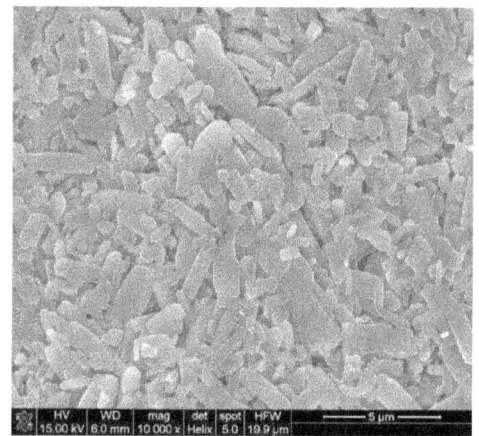

Fig. 11. SEM micrographs of V$_2$O$_5$ samples obtained under different conditions: A – 4 h of sintering in air at 573 K, B – 24 h of sintering in a 10% H$_2$/Ar gas mixture at 573 K.

The typical SEM micrographs of samples sintered at 573 K, either in air or in a 10% H$_2$/Ar gas mixture, are presented in Fig. 11 A, B, respectively. As can be seen, the samples are poly-dispersed, with elongated grains in shape. The determined mean grain size values are listed in Table 3. The average EDX analysis carried out during SEM observations confirmed that the samples have the same chemical composition. The samples sintered in a reducing atmosphere had slightly decreased grain size compared to those sintered in oxidizing conditions at the same temperature (573 K).

Table 3 Mean grain size determined for V$_2$O$_5$ samples depending on sintering atmosphere and time.

Sintering conditions	Grain size	
	Length [μm]	Width [μm]
Air atmosphere, 573 K, 4 h	2.4 ± 0.9	1.0 ± 0.4
10% H$_2$/Ar atmosphere, 573 K, 24 h	2.2 ± 0.5	0.8 ± 0.2

Electrical properties

Fig. 12 shows examples of impedance spectra determined at different temperatures, presented on complex plane Z'' vs. Z' (Nyquist plot).

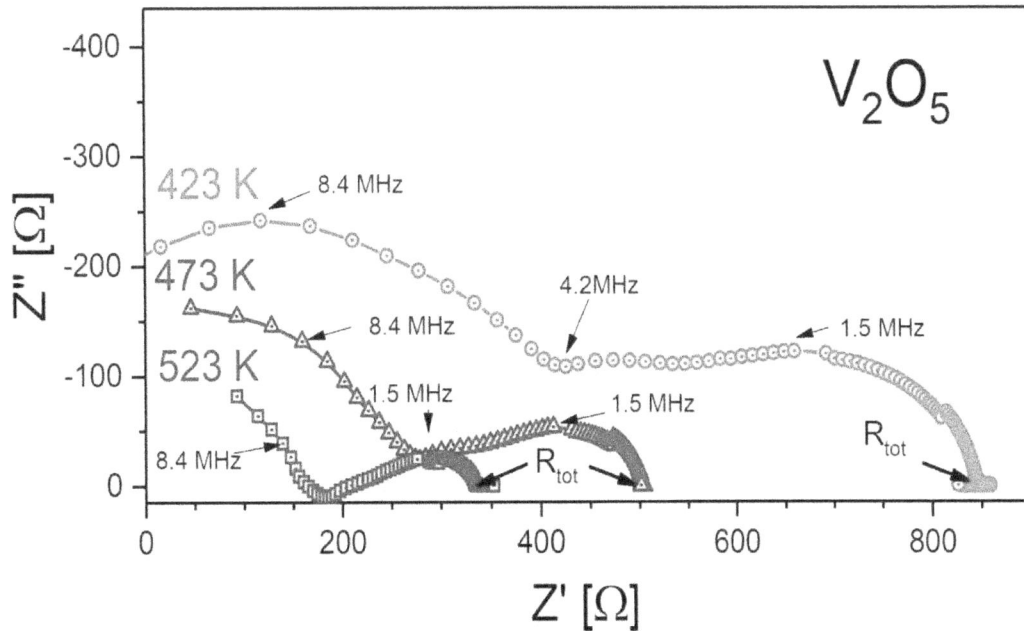

Fig. 12 Nyquist plots obtained for V_2O_5 sintered at different temperatures.

The separation of a high-frequency region corresponding to the bulk properties and an intermediate-frequency region corresponding to the grain boundary properties was difficult due to the partial overlap of the two parts of the spectra. However, it was easy to determine total resistivity (R_{tot}).

Study of MIT occurrence

Fig. 13 illustrates the electrical resistance of V_2O_5 as a function temperature. Such a dependence is typical to semiconductors (R decreases with temperature). No abrupt change in resistance was observed, especially at a temperature close to 530 K, which is the temperature at which MIT was believed to occur in nano-structured V_2O_5 materials (such as thin films [3, 44, 45] and nanowires [46]). The arrow indicates the point at which the MIT might be expected (according to literature reports).

The effect of the type of atmosphere on the electrical properties of a V_2O_5 thin film is presented in Fig. 14. As in the case of the results displayed in Fig. 13, electrical resistance was measured as a function of temperature in an air atmosphere. The sample exhibited typical semiconductor properties. The atmosphere was then changed from air to 10% H_2/Ar when the temperature reached 723 K. The observed abrupt change in electrical resistance was similar to what is associated with the MIT phenomenon.

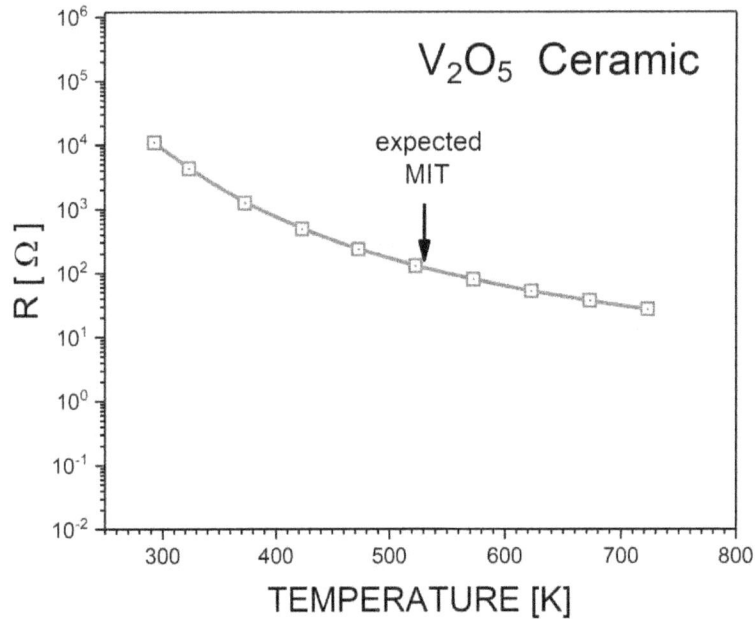

Fig. 13 Electrical resistance of V_2O_5 as a function of temperature. The arrow indicates the point at which an MIT is to be expected according to literature reports.

Moreover, the sample exhibited metallic properties in the 10% H_2/Ar atmosphere, with a temperature coefficient of resistance (*TCR*) of $1.7 \cdot 10^{-4}\,K^{-1}$ – a value typical of metals.

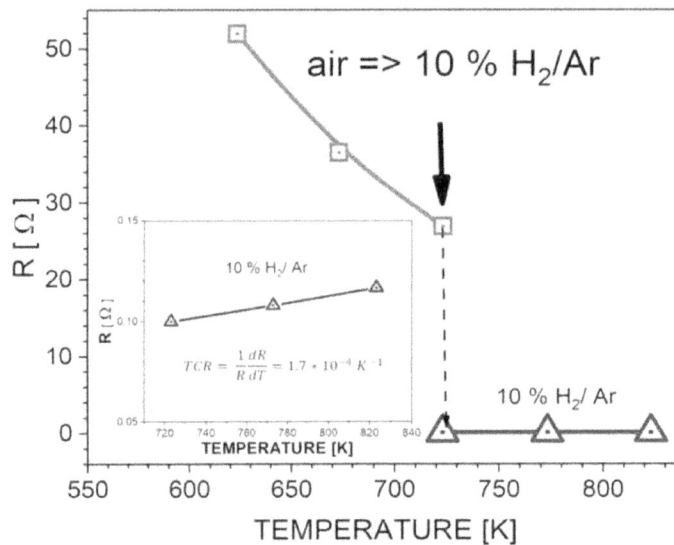

Fig. 14 Electrical resistance of V_2O_5 as a function of temperature in air and after replacing air with an atmosphere consisting of a 10% H_2/Ar gas mixture. Inset: Resistance in the latter atmosphere shown using a different scale.

The observed change in the resistance of the studied sample at 723 K resulted from a chemical reaction, i.e. the reduction of V_2O_5 to V_2O_3 in a strongly reducing gas atmosphere.

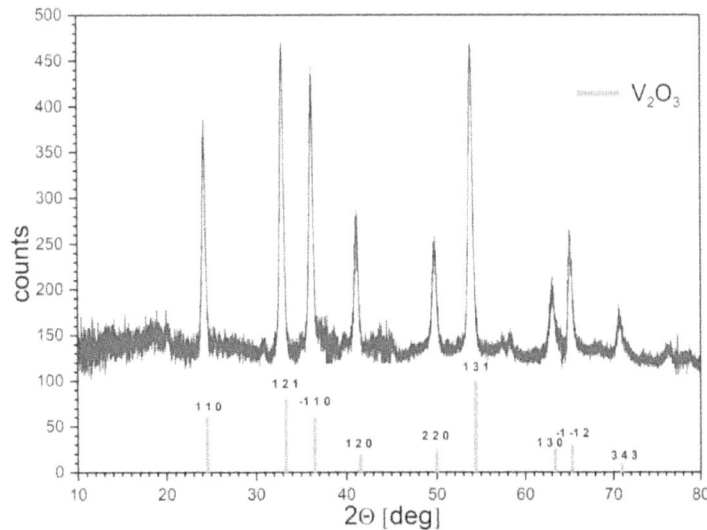

Fig. 15 X-ray diffraction patterns recorded for a V_2O_5 thin film after sintering in a 10% H_2/Ar gas mixture at 723 K.

An XRD analysis (Fig. 15) showed that the V_2O_5 sample had been reduced to the V_2O_3 metallic phase (T_{MIT} = 163 K for V_2O_3). Consequently, the plot shown in Fig. 14 does not represent the MIT phenomenon – during a typical MIT, the material does not change its chemical composition, but it can only undergo a polymorphic phase transition, if at all.

6.2 Study of MIT in V_2O_5 thin films

This section describes the study of the electrical properties of vanadium pentoxide (V_2O_5) thin films.

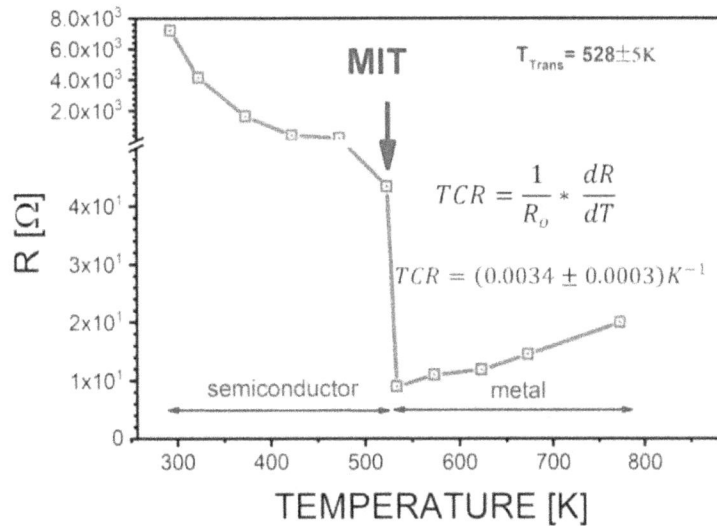

Fig. 16 Dependence of electrical resistance of a V_2O_5 thin film on temperature. The arrow indicates the point at which an MIT takes place.

The results are taken from own studies described in detail in a paper that will be published elsewhere [46]. X-ray diffraction (XRD) and scanning electron microscopy (SEM) were used to determine the structure and phase composition. Electrical properties were analyzed as a function of temperature and oxygen activity by means of electrochemical impedance spectroscopy (EIS).

Fig. 16 illustrates the electrical resistance of V_2O_5 as a function of temperature. Below 500 K, the dependence was typical of semiconductors. At 528 K, an abrupt change in resistance was observed. Above 528 K, the material exhibited metallic behaviour – its resistance increased with temperature. The estimated temperature coefficient of resistance was $3.4 \cdot 10^{-3}$ K^{-1}, which is typical of metals.

These data prove the occurrence of the MIT in the studied thin films. Other authors had also observed this for nano-structured V_2O_5 materials (such as thin films [3, 44] and nanowires [45]). On the other hand, no MIT is observed for ceramic V_2O_5 [40].

7. MIT – APPLICATIONS

The optical properties of vanadium oxides can change under exposure to certain external stimuli in the form of photon radiation (photochromic), a change in temperature (thermochromics), and a voltage pulse (electrochromic); the change becomes discontinuous during an MIT. Such properties can be utilized to prepare coatings for energy-efficient "smart windows" [48] as well as electrical and optical switching devices [49]. Thin films of VO_2 and V_2O_3 have been found to show desirable thermochromics in the infrared region [50, 51]. While maintaining transparency vs. visible light, a smart window modulates infrared irradiation from a low-temperature transparent state to a high-temperature opaque state [52]. Two oxides – VO_2 and V_2O_5 – can change their optical properties in a persistent and reversible way in response to voltage [53]. V_2O_5 exhibits exceptional electrochromic behaviour, as it is characterized by both anodic and cathodic electrochromism, unlike VO_2, which only exhibits the anodic one. These electrochromic materials have four main applications: information displays, variable-reflectance mirrors, smart windows and variable-emittance surfaces. The main applications of the MIT phenomenon in vanadium dioxide thin films are electrical and optical switches.

The field of oxide electronics, i.e. transistor structures based on materials that undergo an MIT, dates back to 1996, when Zhou et al. [54, 55] published a paper in which they presented the idea of a field-effect transistor (FET) based on a hypothetical molecular layer undergoing a Mott transition [8, 54]. Such a device has been called a Mott-FET or MTFET – Mott-transition field-effect transistor. It should be noted that IBM, for example, adopted three main approaches to the development of the so-called "beyond silicon" electronics [53], namely quantum computers, molecular electronics and a thin film MTFET. The latter is considered to be the most promising, because it is closest to the existing circuits and architectures [56].

Vanadium dioxide is currently considered the most suitable material for an MTFET implementation. It should be noted that a simpler material exhibiting the Mott MIT, such as heavily doped silicon, for which this transition occurs at a free charge carrier density of nc ~ 3.5×10^{18} cm^{-3} [25], would seem a better choice for this purpose. However, the Mott transition in doped Si is the second-order phase transition, and it is therefore not accompanied by a jump in conductivity [49]. On the other hand, in vanadium dioxide, the change in conductivity at the transition temperature ($T_{Trans}/T_{MIT} = 340$ K [25]) is as high as 4.5 orders of magnitude (Fig. 3).

One of the most spectacular applications of vanadium oxides materials that undergo the MIT is as memristors. Memristors are 'resistors with memory'. They are well-known as the fourth fundamental passive circuit element (apart from resistors, capacitors and inductors), which were theoretically predicted by L.O. Chua in 1971 [57].

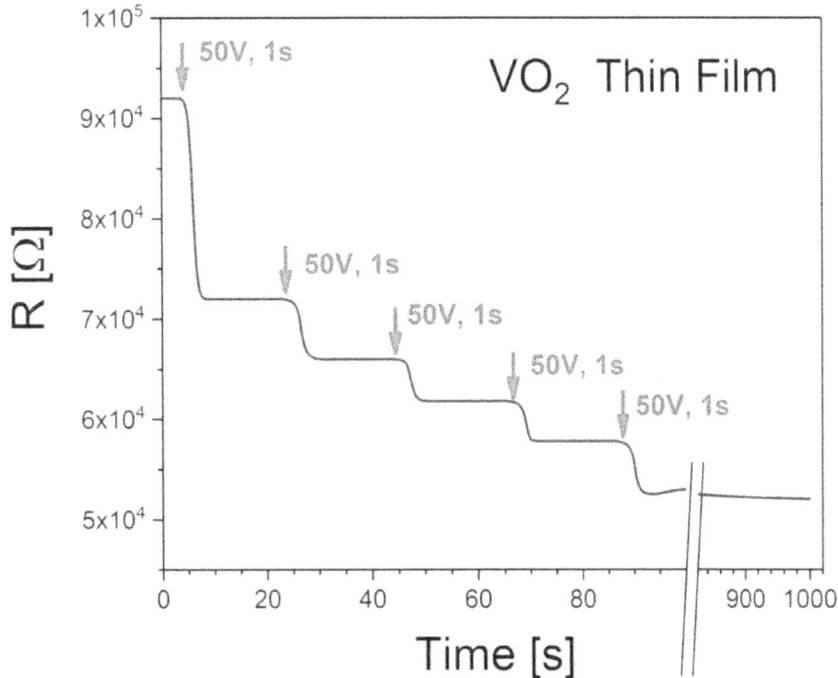

Fig. 17 Changes in the resistance of VO_2 thin films induced by pulse triggers [56-58].

A feature of a memristor is that the resistance of the device depends not only on the instantaneous value of the applied voltage, but rather on the entire dynamic history of charge flowing in the system (Fig. 17). Since memristors can achieve both high integration density and low switching power consumption, they have proved to be very attractive components for next-generation memory technologies [59]. They play significant roles in developing neuromorphic circuits, spintronics, ultra-dense information storage and other applications [60, 61, 62].

VO_2 is one of the most extensively used materials owing to the fact that its T_{MIT} is close to room temperature and the possibility of controlling its MIT via the applied current [63] and electric field [64].

The hysteresis of MITs, presented in Fig. 9, represents the memory aspect of the memristor. This memory is persistent between subsequent ramp voltage pulses (ca. 50 V during 1 s), and non-ohmic behaviour is observed (each pulse triggers the transition to a new resistivity level – Fig. 17) [64, 65]. The changes in resistivity induced by the voltage pulse show very high reproducibility [66]. It has recently been shown that the same effect which can lead to memory-resistance in VO_2 may also lead to memory-capacitance and memory-inductance [67]. In VO_2, the percolative nature is interpreted to be a consequence of the phase-transition phenomenon that accompanies the MIT, leading to an inhomogeneous mixture of insulating and metallic domains at the nanoscale. This inhomogeneity results in dramatic changes in the resistance and the dielectric constant of the material during the phase transition [59, 67].

Fig. 18 shows the areas of the most frequent technical applications of the MIT in vanadium oxides [17, 68-96].

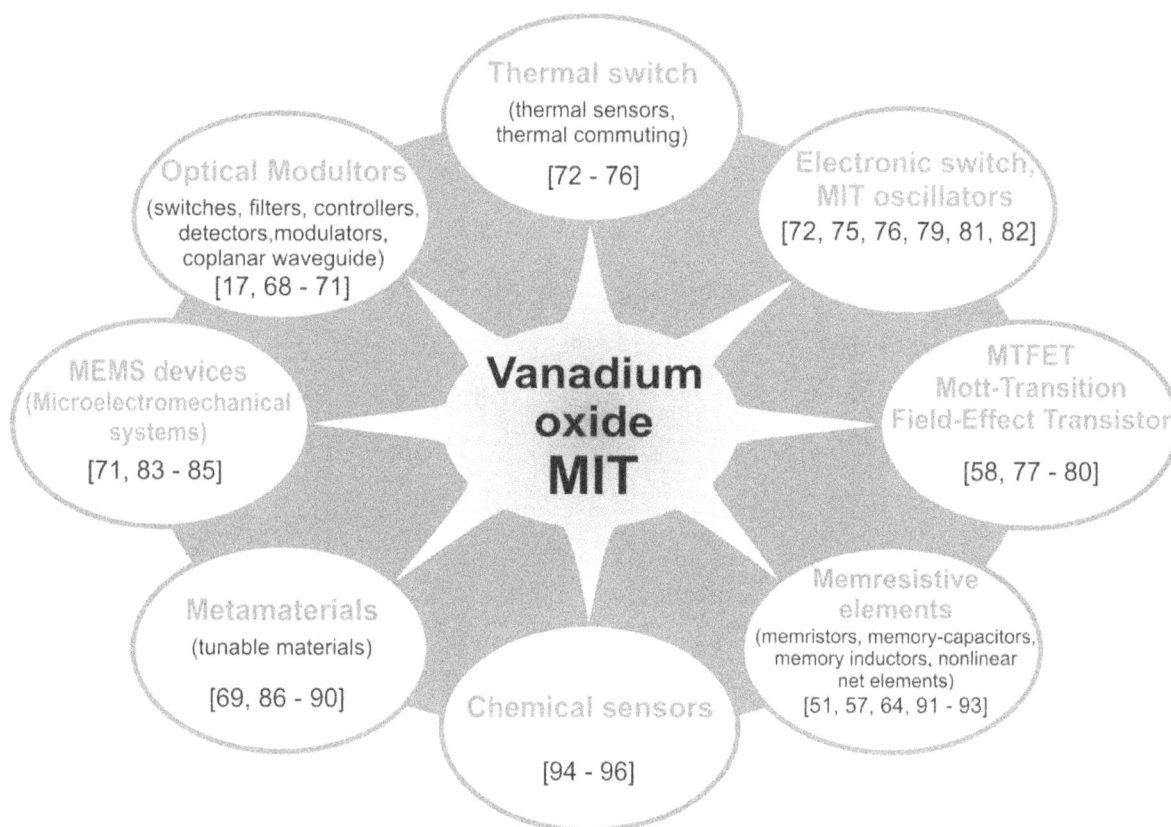

Fig. 18 Devices utilizing MITs in vanadium oxides for oxide electronics

8. CONCLUSIONS

The Mott-Hubbard model of the metal-insulator transition (MIT) was described. The dependence of the temperature at which the metal-insulator transition occurs (T_{MIT}) versus the chemical composition of vanadium oxides was discussed. A detailed description of the MIT in VO_2 and V_2O_5 and its practical applications was presented.

No MIT was observed for the V_2O_5 phase. Although an MIT-like phenomenon was observed after changing the surrounding gas phase from air to 10% H_2/Ar, this process could not have been interpreted as an MIT, but was instead attributable to a chemical reaction (i.e. reduction). XRD analysis revealed that the metallic phase of V_2O_3 was the product of this reaction.

On the other hand, an abrupt change in resistivity was observed at 528 K in the case of a vanadium pentoxide thin film. This phenomenon was interpreted to be a metal-insulator transition. Above 528 K, it would be possible to interpret the impedance spectra using an equivalent circuit composed of a resistor (R), inductor (L), and a constant phase element (CPE) connected in series. The effect of the CPE on the impedance spectra would have been very minor. Above 528 K, the studied sample exhibited metallic properties.

REFERENCES

1. M. Osmolovskaya, I.V. Murin, V.M. Smirnov, M.G. Osmolovsky, *Synthesis of vanadium dioxide thin films and nanoparticles: A brief review*, Rev.Adv. Mater. Sci. 36 (2014) 70-743
2. A. Pergament, G. Stefanovich, V. Andreeev, *Comment om 'Metal-insulator transition without phase transition in V2O5 film'*, Appl. Phys.Lett. 102 (2013) 176101.

3. M. Kang, I. Kim, S. Kim, H.Y. Park, *Metal-insulator transition without structural phase transition in V_2O_5 film*, Appl. Phys. Lett 98(2011) 131907-131916.

4. A. Kämper, I. Hahndorf, M. Baerns, *A molecular mechanics study of the adsorption of ethane and propane on V_2O_5 (001) surfaces with oxygen vacancies*, Top. Catal. 11-12 (2000) 77-84

5. P. Blum, H. Niehus, C. Hucho, R. Fortrie, M.V. Ganduglia-Pirovano, J. Sauer, S. Shaikhutdinov, H-J. Freund, *Surface metal-insulator transition on a vanadium pentoxide (001) single crystal*, Phys. Rev. Lett. 99 (2007) 226103.

6. P.P. Boriskov, A.A. Velichko, A.L. Pergament, G.B. Stefanovich, D.G. Stefanovich, *The effect of electric field on metal-insulator phase transition in vanadium dioxide* Technical Physics Letters 17 (2002) 406-408.

7. N.F. Mott, *Metal-insulator transition*, Rev. Mod.Phys 40, 677-683.

8. M. Imada, A. Fujimori, Y. Tojura, *Metal-insulator transitions*, Rev. Mod. Phys 70 (1998) 1039-1263.

9. J. Hubbard, *Electron correlations in narrow energy bands*. Proceedings of the Royal Society of London. **276** (1963): 238-257.

10. Y. Tokura, *Correlated-electron physics in transition-metal oxides*, *Phys.Today* 56 (2003) 50-55.

11. A. Assmitsu, Y. Tomioka, H. Kuwahara, Y. Tokura, *Current switching of resistive states in magnetoresistive manganites*, Nature 388 (1997) 50-52.

12. K. Miyano, T. Tanaka, Y. Tomioka, Y. Tokura, *Photoinduced insulator–to-metal transition in a perovskite manganite*, Phys.Rev. Lett.78 (1997) 4257-4260.

13. M. Fiebig, K. Miyano, Y. Tomioka, Y. Tokura, *Visualization of the local insulator-metal transition in $Pr_{0.7}Ca_{0.3}MnO_3$*, Science 280 (1998) 1925-1928.

14. J. Cao, E. Ertekin, V. Srinivasan, W. Fan, S. Huang, H. Zheng, J.W. Wim, D.R. Khanal, D.F. Ogietree, J.C. Grossman, J. Wu, *Strain engineering and one-dimensional organization of metal-insulator domains in single-crystal vanadium dioxide beams*, Nat. Nanotechnol. 4 (2009) 732-737.

15. E. Strelcov, Y. Lilach, A. Kolmakov, *Gas sensor based on metal-insulator transition in VO_2 nanowire* 1411.

16. F. Gebhard, The Mott metal-insulator transition, Models and methods, Springer Tracts in Modern Physics, Berlin 1997, ISBN 978-3-540-14858-6.

17. Z. Yang, C. Ko, S. Ramanathan, *Oxide Electronics Utilizing Ultrafast Metal-Insulator Transitions*, Ann. Rev.Res. 41 (2011) 337-367.

18. P.W. Anderson, *Absence of diffusion in certain random lattices*, Phys.Rev. 109 (1958) 1492-1505].

19. T.F. Resenbaum, K. Andres, G. A. Thomas, R. N. Bhatt, *Sharp metal-insulator transition in a random solid*, Phys. Rev. Lett. 45 (1980) 1723-1726.

20. F.J. Morin, *Oxides which show a metal-to-insulator transition at the Neel temperature*, Phys. Rev. Lett. 3 (1959) 34-36.

21. I.G. Austin, *Effect of pressure on the metal-to-insulator transition in vanadium trioxide*, Phil. Mag. 7 (1962) 961-967.

22. K. Sakata, T. Sakata, *Study of semiconductor to metal transition*, Trans. Nat. Res. Inst. Metals 10 (1968) 9-14.

23. R.E. Loechman, C.N. R. Rao, J.M. Honig, *Crystallography and defect chemistry of solid solutions of vanadium and titanium oxides*, J.Phys.Chem 73 (1969) 1781-1784.

24. M.D. Banus, T.B. Reed, *The chemistry of extended defects in non-metallic solids*, L.Eyring and M. O; Keeffe, ed., North Holland, Amsterdam 1970 p. 488-522.

25. F.J. Morin, *Oxides of the 3d transition metals*, Bell. Syst. Tech.J. 37 (1958) 1047-1084.

26. N. F. Mott, *Metal-insulator transition, 2nd ed.* London: Taylor and Francis, 1990.

27. A. Pergament, G. Stefanovich, N. Kuldin, and A. Velichko, *On the problem of metal–insulator transitions in vanadium oxides*, ISRN Condensed Matter Physics, vol. 2013, 2013. Available: http://dx.doi.org/10.1155/2013/960627.

28. A. L. Pergament, G. B. Stefanovich, A. A. Velichko, and S. D. Khanin, *Electronic Switching and Metal-insulator transitions in compounds of transition metals,*" in Condensed Matter at the Leading Edge. Nova Science Publishers, 2006, 1-67.

29. K. Schneider Part II, this issue.

30. K. Schneider Part IV, this issue.

31. Y.Ueda, K. Kosuge, and S. Kachi, *Phase diagram and some physical properties of V_2O_{3+x} ($0 \leq x \leq 0.080$),* J.Solid State 31 (1980) 171-188.

32. I.G. Austin, and C.E. Turner, *The nature of the metallic state in V_2O_3 and related oxides,* Philos.Mag. 19 (1969) 939-949.

33. Ch. Lamsal, N.M. Ravindra, *Optical properties of vanadium oxides an analysis,* J. Mater. Sci, 48 (2013) 6341-6351.

34. A. Pergament, A. Crunteanu, A. Beaumont, G. Stefanovich, A. Velichko, *Vanadium dioxide: Metal-insulator transition, electrical switching and oscillations. A review of state and recent progress,* Energy Materials and Nanotechnology (EMN) Meeting on Computation and Theory, Istanbul (Turkey) 9-12. 11 2015.

35. H. J. Schlag and W. Scherber, *New sputter process for VO_2 thin films and examination with MIS-elements and C–V-measurements,* Thin Solid Films 366 (2000) 28.

36. X.Chen, J. Dai, *Optical switch low-phase transition temperature based on thin nanocrystalline VO_x film,* Optic 121 (2010) 1529-1533.

37. T. Wu, C.J. Patridge, S. Banerjee, G. Sambandamurthy, *Metal-insulator in individual nanowires of doped V_2O_5,* American Physical Society, APS Meeting, March 15-19 2010, abstract #V16.007.

38. W-T. Liu, J. Cao, W. Fan, Z. Hao, M.C. Martin, Y.R. Shen, J. Wu, F. Wang, *Intrinsic optical properties of vanadium dioxide near the Insulator-Metal transition.* Nano Letters, 11 (2011) 466-470.

39. A.L. Pergament, *Metal-insulator transition temperatures and excitonic phases in vanadium oxides.* ISRN Condensed matter Physics, ISRN Condensed Matter Physics, 2011 (2011) Article ID 605913.

40. K. Schneider, K. Kluczewska, M. Dziubaniuk, J. Wyrwa, *Impedance spectroscopy of vanadium pentoxide,* 6th European Young Engineering Conference, EYEC Monograph, Warsaw (2018), M. Nowak, Ed. pp.227-240, ISBN 978-83.936575.5.1.

41. J.D. Hanawalt., H.W. Rinn, L.K. Frevel, *Chemical analysis by X-ray diffraction,* Ind. Eng. Chem. Anal. Ed. 10, 1938, 457-512.

42. R.W.G. Wyckoff., Crystal structures, 2nd ed., Interscience, New York, 1964 v.2.

43. R. Enjalbert., J. Galy, *A refinement of the structure V_2O_5,* Acta Crystallogr. Sect. C: R. Cryst. Struct. Commun. 1986, 42, 1467-1469.

44. E.E. Chain, *Optical properties of vanadium dioxide and vanadium pentoxide thin films.* Applied Optics, 30 (1991) 2782-2787.

45. P. Kiri, G. Hyett, R. Binions, *Solid state thermochromic materials nanocomposite thin films,* Adv.Mater.Lett. 1 (2010) 86-105.

46. K. Schneider, M. Dziubaniuk, J. Wyrwa, *Impedance spectroscopy of vanadium pentoxide thin films,* Journal of Electronic Materials (pending publication).

47. W. Lambrecht, B. Djafari-Rouhani, M. Lannoo, J. Vennik, *The energy band structure of V_2O_5. I. Theoretical approach and band calculations,* J. Phys.C:Solid State Phys. 13 (1980) 2485-2501.

48. C.G. Granqvist, *Spectrally Selective Coatings for Energy Efficiency and Solar Applications.* Physica Scripta, 32 (1985) 401-407.

49. V. Simic-Milosevic, N. Nilius, H-P. Rust, H-J Freund, *Local band gap modulations in non-stoichiometric V_2O_3 films probed by scanning tunneling spectroscopy,* Physical Review B 77(2008)125112-125117.

50. J.B. Kana Kana, J.M. Ndjaka, P. Owono Ateba, B.D. Ngom, N. Manyala, O. Nemraoui, A.C. Beye, M. Maaza, *Thermochromic VO₂ thin films synthesized by rf-inverted cylindrical magnetron sputtering.* Applied Surface Science, 254 (2008) 3959-3963.

51. M.S. Thomas, J.F. DeNatale, P.J. Hood, *High–temperature optical properties of thin–film vanadium oxides – VO₂ and V₂O₃.* Materials Research Society Symposium Proceedings, 479 (1997) 161-166.

52. Z. Zhang, Y. Gao, H. Luo, L. Kang, Z. Chen, J. Du, M. Kanehira, Y. Zhang, Z.L. Wang, *Solution-based fabrication of vanadium dioxide on F: SnO₂ substrates with largely enhanced thermochromism and low-emissivity for energy-saving applications,* Energy & Environmental Science, 4 (2011) 4290-4297.

53. C.G. Granqvist, Handbook of Inorganic Electrochromic Materials (1995) Amsterdam, Holland: Elsevier Science.

54. C. Zhou, D. M. Newns, J. A. Misewich, and P. C. Pattnaik, *A field effect transistor based on the Mott transition in a molecular layer*, Appl. Phys. Lett. 70 (1997) 598-600.

55. D.M. Newns, J.A. Misewich, and C. Zhou, *Nanoscale Mott transition molecular field effect transistor*, U.S. Patent YO996-06, 1996.

56. E. J. Lerner, *The end of the road for Moore's law*, Think Research IBM Thomas Watson Research Center, USA,4 (1999) 7-11.

57. L.O. Chua, *Memristor- the missing circuit element,* IEEE Trans. Circuit Theory 18 (1971) 507-519.

58. A.L. Pergament, G.B. Stefanovich, A.A. Velichko, *Oxide electronics and vanadium dioxide perspective: A review,* J. Selected Topics in Nano-Electronics and Computing, 1 (2013) 24-43.

59. J.J. Yang, M.D. Pickett, X. Li, A.A. Ohlberg Douglas, D.R. Stewart, R.S. Wiliams, *Memristive switching mechanism of metal/oxide/metal nanodevices,* Nat. Nanotechnol. 3 (2008) 429-433.

60. Y.V. Pershin, M. Di Ventra, *Spin memristive systems: Spin memory effects in semiconductor spintronics,* Phys.Rev. B 78 (2008) 159905.

61. M. Son, X. Liu, S.M. Sadaf, D. Lee, S. Park, W. Lee, S. Kim, J. Park, J. Shin, S. Jung, M-H. Ham, H. Hwang, *Self-selective characteristics of nanoscale VOₓ devices for high-density ReRAM applications,* IEEE Electr. Dev. Lett. 33 (2012) 718-720.

62. P. Ben-Abdullah, *Thermal memristor and neuromorphic networks for manipulating heat flow,* AIP Adv. 7 (2017) 065002, 1-6.

63. H-T Kim, B-G. Chae, D-H. Youn, S-L. Maeng, G. Kim, K-Y. Kang, Y-S. Lim, *Comparative analysis of VO₂ thin films prepared on sapphire and SiO₂/Si substrates by the sol-gel process, Jap.* J. Appl. Phys. 46 (2007) 738-741.

64. T. Driscoll, H-T. Kim, B-G. Chae, M. Di Ventra, D.N. Basov, *Phase-transition driven mristive system,* Appl. Phys. Lett. 95 (2009) 043503, 1-3.

65. B.-G. Chae, H-T. Kim, S.J. Jun, *Charcteristics of W and Ti-doped VO₂ thin films prepared by sol-gel method,* Electrochim. Solid State Lett. 11 (2008) D53.D55.

66. S-H. Bae, S. Lee, H. Koo, L. Bin, B-H Jo, C. Park, Z.L Wang, *The memristive properties of a single VO₂ nanowire with switching controlled by self-heating,* Adv. Mater 6 (2013) 1-6.

67. M. Di Ventra, Y.V. Pershin, L.O. Chua, *Circuit elements with memristors, mamcapacitors and meminductors,* Proc IEEE 97 (2009) 1717, Condensed Matter, Mesoscale and Nanoscale Physics.

68. R.M. Briggs, I.M. Pryce, H.A. Atwater, *Compact silicon photonic wave-guide modulator based on the vanadium dioxide metal-insulator phase transition*, Opt. Express 18 (2010) 11192-201

69. F. Cilento, C. Giannetti, G. Ferrini, S. Dal Conte, T. Sala, G. Coslovich, M. Rini, A. Cavalleri, F. Parmigiani, *Ultrafast insulator-to-metal phase transition as a switch to measure the spectrogram of a supercontinuum light pulse.* Appl. Phys. Lett. 96 (2010) 021102

70. J. Liang, J. Li, L. Hou, X. Liu, *Tunable metal-insulator properties of vanadium oxide thin films fabricated by rapid thermal annealing,* ECS J. Solid State Sci. Technol. 5 (1016) 293-298

71. K.C. Pan, W. Wang, E. Shin, K. Freeman, G. Subramanyam, Trans. Electron Devices 3 (2015) 1-6

72. C. Ruzmetov, G.Gopalakrishnan, J.D. Deng,V. Narayanamurti, S. Ramanathan, *Electrical triggering of metal-insulator transition in nanoscale vanadium oxide junctions*. J. Appl. Phys. 106 (2009) 083702.

73. B.J. Kim, Y.W. Lee, S. Choi, S.J. Yun, H.T. Kim, *VO₂ thin-film varistor based on metal-insulator transition.* IEEE Electron Device Lett. 31 (2010) 14-16.

74. Z. Yang, C. Ko, V. Balakrishnan, G. Gopalakrishnan, S. Ramanathan, *Dielectric and carrier transport properties of vanadium dioxide thin films across the phase transition utilizing gated capacitor devices*. Phys. Rev. B 82 (2010) 205101.

75. J. Kim, C. Ko, A. Frenzel, S. Ramanathan, J.E. Hoffman, *Nanoscale imaging and control of resistance switching in VO₂ at room temperature*, Appl. Phys. Lett. 96 (2010) 213106.

76. C. Ko, S. Ramanathan, *Observation of electric field-assisted phase transition in thin film vanadium oxide in a metal-oxide-semiconductor device geometry*, Appl. Phys. Lett. 93 (2008) 252101.

77. Y. Zhou, S. Ramanathan, *Correlated electron materials and field effect transistors for logic: A review,* Critical Review in Solid State and Mater. Sci. 38 (2013) 286-316.

78. J. Jeong, N. Aetukuri, T. Graf, T.D. Schladt, M.G. Semant, S.S.P. Parkin, *Supression of metal-insulator transition in VO₂ by electric field-induced oxygen vacancy formation,* Science 339 (2013) 1402-1405.

79. D. Ruzmetov, G. Gopalakrishnan, C. Ko, V. Narayanamurti, S. Ramanathan, *Three-terminal field effect devices utilizing thin film vanadium oxide as the channel*. J. Appl. Phys. 107(2010) 114516.

80. S. Hormoz, S. Ramanathan, Limits *on vanadium oxide Mott metal-insulator transition field-effect transistors*, Solid-State Electron. 54 (2010) 654–59.

81. H.T. Kim, B.J. Kim, S.Choi, B.G. Chae, Y.W. Lee, T. Driscoll, M.M. Qazilbash, D.N. Basov, *Electrical oscillations induced by the metal insulator transition in VO₂.* J. Appl. Phys. 107 (2010) 023702.

82. Y.W. Lee, B.J. Kim, J.W. Lim, S.J. Yun, S. Choi, B.G. Chae, G. Kim, H.T. Kim, *Metal-insulator transition-induced electrical oscillation in vanadium dioxide thin film*. Appl. Phys. Lett. 92 (2008) 162903.

83. R. Cabrera, E. Marced, N. Sepulveda, *Performance of electro-thermally driven VO₂–based MEMS actuators*, J. Microelectromechanical Systems, 23 (2014) No. 1 Feb 2014 K

84. F. Niklaus, C. Vieder, H. Jacobsen, *MEMS-based uncooled infrared bolometer arrays- A review*, MEMS/MOEMS Technologies and Applications III, ed. J-C. Chiao, X. Chen, Z. Zhou, X. Li, Proc. of SPIE v. 6836 (2007) 68360D-1.

85. R.B. Darling, S. Iwanaga, *Structure, properties and MEMS and microelectronic applications of vanadium oxides*, Sadhana 34 (2009) 531.542.

86. M.J. Dicken, K. Aydin, I.M. Pryce, L.A. Sweatlock, E.M. Boyd, S. Walavalkar, J. Ma, H.A. Atwater, *Frequency tunable near-infrared metamaterials based on VO₂ phase transition*. Opt. Express 17 (2009) 18330-18339.

87. T. Driscoll, H.T. Kim, B.G. Chae, B.J.Kim, Y.W. Lee, N.M. Jokerst, S. Palit, D.R. Smith, M. Di Ventra, D.N. Basov, *Memory metamaterials,* Science 325 (2009) 1518- 1521.

88. T. Driscoll, S. Palit, M.M. Qazilbash, M. Brehm, F. Keilmann, B-G. Chae, S-J. Yun, H-T. Kim, S.Y. Cho, N.M. Jokerst, D.R. Smith, D.N. Basov, *Dynamic tuning of an infrared hybrid-metamaterial resonance using vanadium dioxide*. Appl. Phys. Lett. 93 (2008) 024101.

89. Q.Y. Wen, H.W. Zhang, Q.H. Yang, Y.S. Xie, K. Chen, Y.L. Liu, 2010. *Terahertz metamaterials with VO₂ cut-wires for thermal tunability*. Appl. Phys. Lett. 97 (2010) 021111.

90. M. Seo, J. Kyoung, H. Park, S. Koo, H.S. Kim, H. Barnien, B.J. Kim, J.H. Choe, Y.W. Ahn, H-T. Kim, N. Park, Q-H. Park, K. Ahn, D-S. Kim, *Active terahertz nanoantennas based on VO₂ phase transition.* Nano Lett. 10 (2010) 2064–2068.

91. Driscoll T, Kim HT, Chae BG, Di Ventra M, Basov DN. 2009. Phase-transition driven memristive system. Appl. Phys. Lett. 95:043503.

92. D.B. Strukov, G.S. Snider, D.R. Stewart, R.S. Williams, *The missing memristor found,* Nature 453 (2008) 80-83.

93. S.H. Jo, T. Chang, I. Ebong, B.B. Bhadviya, P. Mazumder, W. Lu, *Nanoscale memristor device as synapse in neuromorphic systems.* Nano Lett. 10 (2010) 1297-1301.

94. E. Strelcov, Y. Lilach, A. Kolmakov, *Gas sensor based on metal-insulator transition in VO_2 nanowire thermistor,* Nano Lett. 9 (2009) 2322-2326.

95. J.M. Baik, M.H. Kim, C. Larson, C.T. Yavuz, G.D. Stucky, A.M. Wodtke, M. Moskovits, *Pd-sensitized single vanadium oxide nanowires: highly responsive hydrogen sensing based on the metal-insulator transition.* Nano Lett. 9 (2009) 3980-3984.

96. K. Schneider, M. Lubecka, A. Czapla, V_2O_5 *thin films for gas sensor applications*, Sensors and Actuators B: Chemical, 236, (2016) 970-977.

IV. Thin films – preparation, properties, applications

Abstract

The article consists of two parts. The first is a review of the techniques used to prepare vanadium oxides thin films. The second part describes the author's experimental works on the deposition of vanadium pentoxide thin film by means of reactive radio frequency sputtering. The structure and morphology of four series of V_2O_5 thin films prepared at different conditions is described.

Key words: Vanadium oxides, thin films, radio-frequency sputtering, structure, morphology

1. INTRODUCTION

Vanadium oxides (mainly VO_2 and V_2O_5), especially in the form of nanostructured thin films, have recently become a subject of extensive studies owing to their fascinating structural, optical and electrical properties. Their wide optical band gap and good thermal and chemical stability make them a promising material in industrial applications. The properties of vanadium oxide thin films depend mostly on the deposition technique and the deposition conditions such as starting materials, deposition temperature, deposition rate, residual gases in the deposition chamber, etc. The present paper includes a literature review on their preparation techniques as well as the author's own studies on the structure and morphology of four series of V_2O_5 thin films prepared at different conditions.

2. PREPARATION OF VANADIUM OXIDE THIN FILMS – LITERATURE REVIEW

Many thin film growth techniques have been adopted to deposit vanadium oxide thin films on various substrates [1-3]. They include chemical vapour deposition (CVD), chemical reaction methods that utilize precursor solutions, electrochemical methods and physical vapour deposition (PVD).

2.1 CVD (chemical vapour deposition) [4-19]

Chemical vapour deposition (CVD) is an industrial process applied to deposit thin films. In a CVD process, the substrate is exposed to one or more volatile precursors, which react on the substrate's surface, producing the desired film. An advantage of this method is the ability to coat large areas. Advanced versions of the CVD, used for preparation of VO_2 and V_2O_5, include PECVD (plasma-enhanced chemical vapour deposition) [4 - 6] and ALD (atomic layer deposition) [1, 7 - 9].

Various vanadium-containing chemicals have been used as volatile precursors, both metal–organic, such as vanadium acetylacetonate – $(C_5H_7O_2)_4V$ [10, 11] – or vanadyl tri-isopropoxide – $VO(OC_3H_7)_3$ [12], and inorganic ones, such as vanadium oxychloride – $VOCl_3$ [13] – or vanadium(III) chloride – VCl_3 [14]. Piccirillo et al. [12] reported an aerosol-assisted variant of the technique (AA CVD), in which the precursors are dispersed in a solvent and an aerosol of the solution is generated ultrasonically. This precursor is transported to the substrate in the form of aerosol droplets, by a carrier gas. In this technique, the precursor does not need to be volatile, but merely soluble in a solvent suitable for aerosol formation.

Losurdo et al. [15] studied nanocrystalline V_2O_5 thin films grown by means of plasma-enhanced chemical vapour deposition. Both the real and imaginary part of the complex dielectric function and, hence, the refractive index and absorption coefficients, were described up to a photon energy of 5 eV, taking into account the anisotropy of vanadium pentoxide and the influence of the films' microstructure on their optical properties. It was observed that the substrate nature has major influence on the dynamics of nucleation from the vapour phase and, therefore, on the optical properties of the obtained layers.

G. Rampelberg et al. [16] prepared thin films of VO_2 by means of the ALD method using a metalorganic and ozone as precursors at 423K. Crystallization of as-sputtered films is very sensitive to oxygen partial pressure when annealing at 723 K. In an inert atmosphere (He), the annealed films remain amorphous. The crystallization of the films takes place at an oxygen partial pressure of 1-10 Pa. At $p(O_2) >35$ Pa, VO_2 transforms to V_6O_{13}.

Musschoot et al. [17] deposited vanadium pentoxide thin films using the ALD method in the 323-473 K temperature range, starting with vanadyl-tri-isopropoxide (VTIP). $VO(OiPr)_3$ as one precursor. The use of water vapour as a second precursor yielded an amorphous film, whereas the application of oxygen in this role lead to the formation of an oriented V_2O_5 film.

When using the same vanadium source, Chen et al. [18] reported the growth of crystalline V_2O_5 at 443-461 K, with VTIP and ozone as precursors. Annealing in He led to oxygen-deficient films, whereas in O_2 all samples eventually crystallized into V_2O_5.

I.I. Kazadojev et al. [19] prepared V_2O_5 films by means of the ALD method using Tetrakis (dimethylamino) vanadium (IV) ($V(NMe_2)_4$) as the vanadium source and either oxygen-argon plasma, oxygen, or water as the co-reagent. Deposition was performed over 400 cycles and a range of 423-573 K, resulting in a number of both stoichiometric and non-stoichiometric vanadium oxides. Electrochemical characterization revealed that the samples deposited at 523 K via a plasma process and post-growth annealing exhibit the best electrochemical properties in terms of current density and charge density, with values of 0.35 mAcm^{-2} and 55 mCcm^{-2}, respectively. Annealing in air at 400°C for both thermal and plasma-assisted methods resulted in the formation of α–V_2O_5.

2.1.1 Chemical reaction methods based on precursor solutions [1, 20-26]

Chemical reaction methods include spray pyrolysis [1, 20, 21] and electrospinning [1, 22 - 24]. In spray pyrolysis, a liquid solution of the precursor is mixed with a carrier gas and expelled as a fine mist via a spray nozzle. The mist condenses on a heated substrate, where it is pyrolyzed [1].

Abbasi et al. [20] deposited V_2O_5 thin films via the spray pyrolysis of an aqueous solution of 0.05 M VCl_3 on preheated glass substrates (573-773 K). Films prepared below 573K exhibited poor adhesion to the substrate surface. The average grain size of the films prepared at 673 K was around 25 nm.

Basha and Akilasundari [21] prepared V_2O_5 thin films via the spray pyrolysis of an aqueous solution of vanadium (III) chloride (VCl_3) on a glass substrate. The films were prepared at temperatures between 473 and 523 K. As prepared, the films exhibited an orthorhombic structure with a preferred orientation along the (0 0 1) direction. The optical band gap was strongly dependent on substrate temperature, and it varied between 2.32 and 2.46 eV.

The standard setup for electrospinning consists of a spinneret, which is a syringe needle connected to a high-voltage (5 to 50 kV) DC power supply. A precursor solution is loaded into the syringe and is extruded from the needle tip at a constant rate by a syringe pump. When a sufficiently high voltage is applied to a liquid droplet, it becomes charged, and the electrostatic repulsion counteracts the surface tension, which causes the droplet to stretch [1]. Electrospinning is a versatile and effective method for manufacturing quasi-one-dimensional nanostructures, such as nanowires, nanoneedles, nanobelts and nanotubes.

Using this method, Ban et al. [22] prepared single-crystalline vanadium oxide nanorods, L. Mai et al. [24] synthesized ultra-long hierarchical V_2O_5 nanowires, while Ostermann et al. [23] manufactured V_2O_5 nanorods on TiO_2 nanofibres.

C. Diaz et al. [25] describe a novel and simple solvent-less process for the preparation of pure V_2O_5 by means of the solid-state pyrolysis of easily prepared macromolecular complexes. The characteristics of the macromolecular complex precursors, such as the nature of the polymer and the metal/polymer ratio, are crucial factors that determine the morphology and particle size of the pyrolytic products. The application of chitosan polymer as a solid-state template can facilitate the formation of single-crystal nanobelt structures. For the pyrolytic products obtained from chitosan·(VCl$_3$)$_y$ (1:5), nanoparticles with a mean diameter of ~15nm are possible. A. Akl [26] prepared vanadium pentoxide (V$_2$O$_5$) thin films by means of spray pyrolysis. The influence of the solution's molarity on the characteristics of V$_2$O$_5$ was investigated. X-ray diffraction analysis (XRD) showed that the films deposited at $\geq$0.1 M were orthorhombic in structure, with the preferred orientation along the $\langle 0\,0\,1 \rangle$ direction. Moreover, crystallinity improved when the solution's molarity had been increased. Microstructural parameters were evaluated using a single-order Voigt profile method. The optical band gap values, determined using a Tauc plot, were found to be 2.50 ± 0.02 and 2.33 ± 0.02 eV for direct and indirect allowed transitions, respectively. In addition, the complex optical constants for the wavelength range of 300-2500 nm were reported. At room temperature, dark conductivity as a function of the solution's molarity was in the range of $5.74 \cdot 10^{-2} \pm 0.03$ to $3.36 \cdot 10^{-1} \pm 0.02$ Ω^{-1} cm^{-1}, whereas at high temperatures, the character of electrical conductivity is determined predominantly by grain boundaries.. The values of activation energy and potential barrier height were 0.156 ± 0.011 and 0.263 ± 0.012 eV, respectively.

2.1.2 Electrochemical methods [27 -29]

An electrochemical method of thin film fabrication is greatly superior to other methods in terms of cost and flexibility. The method can be employed at low temperatures (typically room temperature) and ambient pressure. The growth of the film can be controlled by varying the electrodeposition conditions such as current density, time and electrolyte composition and concentration [27].

Lu et al. [28] prepared a V$_2$O$_5$ thin film via electrodeposition, using a three-electrode cell with an ITO substrate and a 1:1 mixture of deionized water and ethanol, which contained 1 M of VOSO$_4$. Electrodeposition was performed at a potential of -0.7 V relative to that of the reference electrode for 20-60 s. The films deposited for 20 s were smooth and exhibited good adhesion to the substrate.

Rasoulis and Varnardou [29] synthesized a V$_2$O$_5$ thin film using an electrochemical cell consisting of platinum, Ag/Ag/Cl and SnO$_2$ precoated glass substrates as the counter, reference and working electrodes, respectively. The electrolyte was a solution of 0.5 M vanadyl (III) acetylacetonate – VO(acac)$_3$. The growth of the film was carried out at room temperature, with deposition current densities of 0.7-1.3 Acm^{-2} and for a deposition time of 15 min. Structural analysis indicated that increasing the current density and electrolyte concentration enhanced the crystalline quality of the films.

2.1.3 Sol-gel method [1, 2, 30]

The sol-gel process is one of the oldest methods (dating back to the 19th century), and is still widely employed for the deposition of vanadium oxide thin films. Its low cost is its advantage over expensive vacuum techniques. Moreover, this method can be used to deposit films on a large surface area and offers a simple way of metal doping. The sol-gel method is a wet chemistry technique which utilizes a chemical solution containing colloidal precursors (sol) as a starting material. Typical precursors include metal-organics and vanadium chlorides, which undergo hydrolysis and polycondensation reactions to form a colloid. The sol then evolves toward the formation of an inorganic network containing a liquid phase (gel). Vanadium oxide in the gel forms polymers such as oxo- (V-O-V) or hydroxo- (V-OH-V) bridges in solution. These polymers are deposited on a substrate. The drying process then removes the liquid phase, and a porous oxide film forms. During the subsequent thermal annealing, crystallization proceeds, the mechanical properties are enhanced, and a change in porosity is observed.

The precursor sol can be deposited on the substrate by means of either dip-coating or spin–coating. Using the first technique, the substrate is dipped into the sol, and then slowly withdrawn. The gel formed on the substrate is left to dry at room temperature. In order to obtain thicker films, the procedure is repeated two or more times. The spin-coating technique is used to deposit uniform thin films on flat substrates. The solution is placed on the substrate, which is then rotated, allowing the fluid to be spread by the centrifugal force. The thickness of the film may be controlled by adjusting the spin angular velocity or the concentration of the solution.

Wu et al. [31] synthesized vanadium dioxide thin films on mica substrates at different annealing temperatures using a sol-gel method. The films annealed at 773 K exhibited a polycrystalline structure with high crystallinity and compact surface. In other studies, the sol-gel technique was applied to deposit VO_2 thin films on sapphire [32], aluminium [33], silicon [34] and glass [35].

Ningyi et al. [36] used the sol-gel method to prepare highly oriented V_2O_5 thin films on an SiO_2/Si substrate. The films were reduced to VO_2 in air at 673 K and under a pressure of 2 Pa. The process followed the sequence: $V_2O_5 \rightarrow V_3O_7 \rightarrow V_4O_9 \rightarrow V_6O_{13} \rightarrow VO_2$.

Z.S. El Mandouh and M.S. Selim [37] prepared vanadium pentoxide films by means of an inorganic sol-gel method. The sol was obtained by melting V_2O_5 powder at 1073 K. The liquid was then quickly poured into distilled water. Glass substrates were dipped into the obtained sol and then slowly withdrawn. The gel formed on the substrate was left to dry at room temperature for 24 h. Annealing at 573 and 673 K improved the crystallinity of the films and reduced the localized states in the band gap. Optical measurements made it possible to calculate the band gap values of the prepared films, which were found to be 2.49 and 2.42 eV for the untreated films and those annealed at 473 K, respectively. The determined electrical properties of the films (resistivity and Seebeck coefficient) suggested that the films are suitable for application as thermistor devices.

2.2 PVD (physical vapour deposition) [1, 2]

Physical vapour deposition refers to a variety of vacuum deposition methods. Such methods are characterized by a process whereby the deposited material is first transformed from a condensed phase to vapour and then back to a condensed phase, which forms a thin film. The most common PVD processes are evaporation and sputtering. PVD has been used to deposit both vanadium dioxide [2] and vanadium pentoxide [1].

2.2.1 Thermal evaporation [3, 38-41]

M.C. Rao and K.R. Rao [38] prepared V_2O_5 thin films via the thermal evaporation of pure V_2O_5 powder in an evaporation boat heated to 1823 K in a vacuum ($4 \cdot 10^{-4}$ Pa). The diffraction pattern indicated that the deposited film was amorphous. Singh et al. [39] observed a similar result for thin films deposited on a Ni-coated glass via the thermal evaporation of vanadium pentoxide powder in oxygen.

Wu et al. [40] prepared vanadium pentoxide films via the vacuum evaporation of 99.99% pure V_2O_5 powder using a resistively heated molybdenum boat. The as-prepared films had a distinctive yellow appearance, just like a V_2O_5 powder. X-ray diffraction analysis revealed that the films were amorphous and became microcrystalline after annealing at 623 K. The optical absorption spectra of the samples exhibited two distinct regions of behaviour: one at high energy, in which $(\alpha h\nu)^{1/2}$ varied linearly with photon energy ($h\nu$), and another – a low-energy tail.

Kumar et al. [41] prepared V_2O_5 thin films via thermal evaporation in a vacuum coating unit over an amorphous glass substrate. The source material was V_2O_5 powder with 99.998% purity, evaporated from a molybdenum boat. During deposition, the pressure in the chamber was 10^{-4} Pa and the distance between the substrate and the molybdenum boat was maintained at approx. 7 cm. The deposition rate was about 0.25 µm/min. The deposition temperature was found to have a great impact on the optical and structural properties of the investigated films. The films deposited at room temperature had a homogeneous, uniform and smooth texture, but were amorphous in nature. These films remained amorphous even after additional annealing at 573 K. On the other hand, the films deposited at the substrate temperature of 573 K were highly textured and c-axis-oriented, with good crystalline properties. Moreover, the colour of the films changed from pale yellow to light brown to black depending on whether deposition had been performed at room temperature, 573 or 773 K, respectively. The observed decrease in lattice constants with increased deposition temperature was attributed to oxygen loss in the V_2O_5 structure, which usually leads to reduced (001) interplanar distance. AFM analysis showed that for an increased deposition temperature E_g = 2.58 eV.

2.2.2 Electron-beam evaporation [1, 2, 42-44]

In this method, a high-energy electromagnetically focused electron beam bombards a target. The kinetic energy of the electrons is converted into the thermal energy of atoms on the target's surface. This causes either sublimation or the formation of a liquid melt with a substantial vapour pressure. This method allows the precise control of deposition rate and film composition. V_2O_5 films deposited at room temperature using this method were amorphous in nature. However, films deposited at substrate temperatures above 470 K were polycrystalline, with a c-axis texture [42]. The band gap of the deposited films varied between 2.04 and 2.66 eV. The oxygen partial pressure during deposition and the target's temperature were observed to affect the band gaps [42-44]. There are some discrepancies in the reported results concerning the effect of oxygen partial pressure on the band gap – Ramana et al. [43] suggested that the band gap increased together with oxygen partial pressure, whereas Laurenco et al. [44] claimed the opposite to be true.

2.2.3 PLD (pulsed laser deposition) [1, 2, 45-48]

Pulsed laser deposition, PLD, is a technique in which a high-power pulsed laser beam is focused on a target inside a vacuum chamber. Material vaporized from the target is deposited on a substrate as a thin film. The process can occur in ultra-high vacuum or in the presence of a chosen gas, such as oxygen, which is commonly used when depositing vanadium oxides. The partial pressure of oxygen in the chamber is critical in stabilizing the desired phase of the vanadium oxide. Beke et al. [46] synthesized V_2O_5 thin films at 470-520 K using an ArF pulsed laser ($\lambda = 193$ nm, $\tau = 15$ ns) and a bulk V_2O_3 target. XRD analysis revealed the polycrystalline structure of the film. The band-gap energy (E_g) was 2.52 eV. PLD was also used for the deposition of VO_2 thin films [47, 48]. Bore et al. [47] employed a KrF pulsed excimer laser (248 nm) to ablate a vanadium target in an ultra-high vacuum chamber with an Ar/O_2 (10:1) atmosphere under a pressure of 0.02-0.04 Pa and a substrate temperature of 773 K. These experimental conditions were critical in promoting the formation of the pure VO_2 phase. Suh et al. [48] studied the effect of the deposition of VO_2 thin films by means of the PLD technique on the MIT transition.

2.2.4 Magnetron sputtering [1, 49-60-68]

Magnetron sputtering was performed using either direct current (DC) [49-52] or radio frequency (RF) [53-59]. Metallic vanadium [49-57] or V_2O_5 [58] was used as a target. Films prepared at a temperature below 470 K were amorphous, while those obtained at higher temperatures were crystalline.

The simplest form of sputtering involves a diode geometry in which energetic ions (usually argon ions) from gas discharge plasma, bombarding the target that forms the cathode for the discharge. Target atoms bombard the substrate that plays the role of the anode, forming a film.

Depending on the deposition conditions and the applied method, vanadium oxide thin films have different chemical composition (varying ratio V/O), structure, and considerably different structural, optical and electrical properties. In general, vanadium oxide thin films deposited at a relatively low substrate temperature (below 573 K) are amorphous. An increase of the substrate temperature during film deposition was found to promote the loss of oxygen atoms and thereby a change in the stoichiometry of the VO_x film. Moreover, the temperature of crystallization depends on the method of film deposition [61].

Benmoussa et al. [62] prepared V_2O_5 thin films by means of RF sputtering from a V_2O_3 target in an Ar/O_2 gas mixture under a total pressure of 1 Pa and for an oxygen partial pressure ($p(O_2)$) varied from 0 to 20%. All films were identified as V_2O_5. At 673 K and in vacuum, the films underwent reduction to VO_2. The electrical conductivity (E_{act}) of the film grown for a $p(O_2)$ of 5% was equal to 0.032 eV for the low-temperature range (143-250 K), while at higher temperatures (250-420 K) it was 0.101 eV. The small polaron mechanism was postulated due to the existence of two different valence states: V^{5+} and V^{4+}.

In paper [63], vanadium oxide thin films were grown on quartz substrates using the pulsed DC reactive magnetron sputtering technique at room temperature and afterwards annealed under vacuum conditions in the temperature range from 327 to 503 K. The electrical resistance, temperature coefficient of resistance (TCR), optical energy gap and structural properties were investigated. The films consisted of a V_2O_5 phase with nanoscale grains, the mean size of which increased with temperature. Additionally, the post-annealing at 503 K induced the formation of both V_2O_5 and V_4O_9 phases and pinholes on the film's surface. The temperature-dependent variation of electrical resistance indicated two activation energy areas corresponding to two TCR values for the films post-annealed up to 453 K, but only one activation area was found after annealing at 503 K. Analyses of the absorption coefficient versus photon energy revealed a direct forbidden transition. The mean grain size and TCR values increased with post-annealing temperature, whereas the optical energy gap and electrical resistance did not follow this tendency. The evolution of the structure and its relation to the optical energy gap, electrical resistance, activation energy and TCR were discussed in the context of the results obtained in the present study.

3. PREPARATION OF THIN FILMS – OWN STUDIES

RF reactive ion sputtering was the method used in the presented work. A 99.97% pure vanadium target was sputtered in a flow-controlled Ar/O_2 gas atmosphere. The deposition chamber is presented schematically in Fig.1. The vacuum chamber contained two electrodes: a cathode-target and an anode-substrate holder. The substrate holder was equipped with a temperature- controlled heater. Fig.2 shows the applied RF equipment (Department of Electronics, AGH University of Science and Technology, Krakow). The parameters of deposition are listed in Table 1. A detailed description is given elsewhere [61, 69-74].

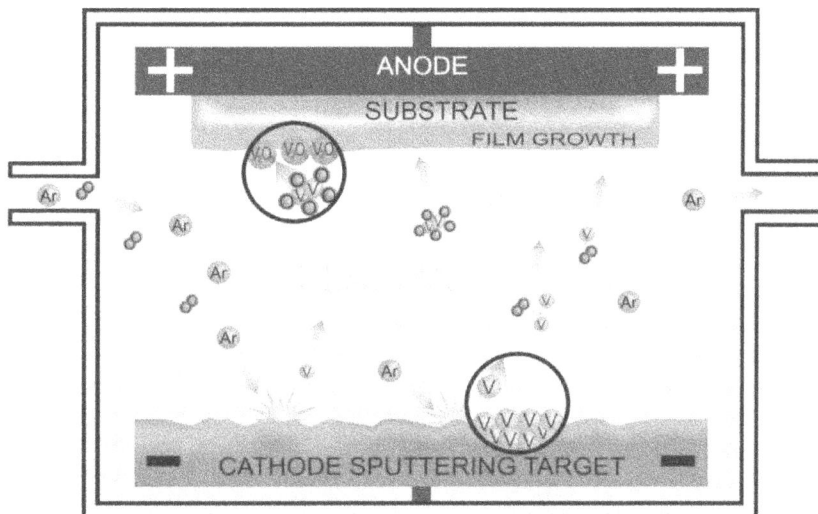

Fig. 1 Schematic view of the RF deposition chamber

Fig. 2 RF thin film deposition equipment

Table 1 Deposition conditions for a series of vanadium oxide thin films and their properties

Deposition conditions & methods	1	2	3
Ar flow [cm^3/s]	6.3	6.7	6.7
O_2 flow [cm^3/s]	0.3	0.7	2.1
Input power [W]	290	280	290
RF voltage (U_{rf}) [V]	1150	1150	1150
Ar/O_2 gas atmosphere pressure [Pa]	$4.1 \cdot 10^{-2}$	$4.3 \cdot 10^{-2}$	$4.7 \cdot 10^{-2}$
Deposition time [min]	240	240	240 360
Thickness [nm]	420 ± 49	421 ± 43	431 ± 28 630 ± 42
Substrate	Corning, Ti, fused SiO_2	Corning, Ti, fused SiO_2,	Corning, Ti, fused SiO_2, BVT
Substrate temperature [K]	298	298	298
Ref.	[64]	[64]	[61, 64]
Description	Ch. 3.1..1	Ch. 3.1..2	Ch 3.1..3
Experimental methods	XRD, SEM EDX, optical	XRD, SEM, EDX	XRD, SEM, EDX, EIS

3.1 Structure and morphology of thin films

Figs 3 and 4 present the X-ray diffraction patterns recorded for as-sputtered thin films with a thickness of 430 nm and 630 nm, respectively. The thicker sample ($d = 630$ nm) was characterized by more developed peaks ($d = 430$ nm). The determined lattice parameters are listed in Table 2 along with literature data taken from refs [74 -77]. As can be seen, these two sets of values are highly consistent. The positions of peaks correspond to those defined in the JCPDS #09-0387 standard. However, differences in relative intensities are observed – for example, the maximum intensity of the 630 nm film's peak in the present work is for (2,0,0), and not for (0,0,1), as in JCPDS. This fact can be explained by the preferred orientation of crystallites in this film.

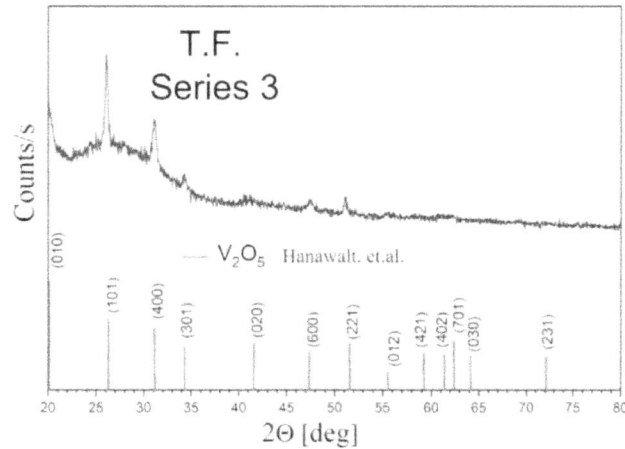

Fig. 3 X-ray diffraction pattern recorded for a Series 3 thin film obtained after 240 min of deposition ($d = 430$ nm); reference lines taken from Ref. [74]

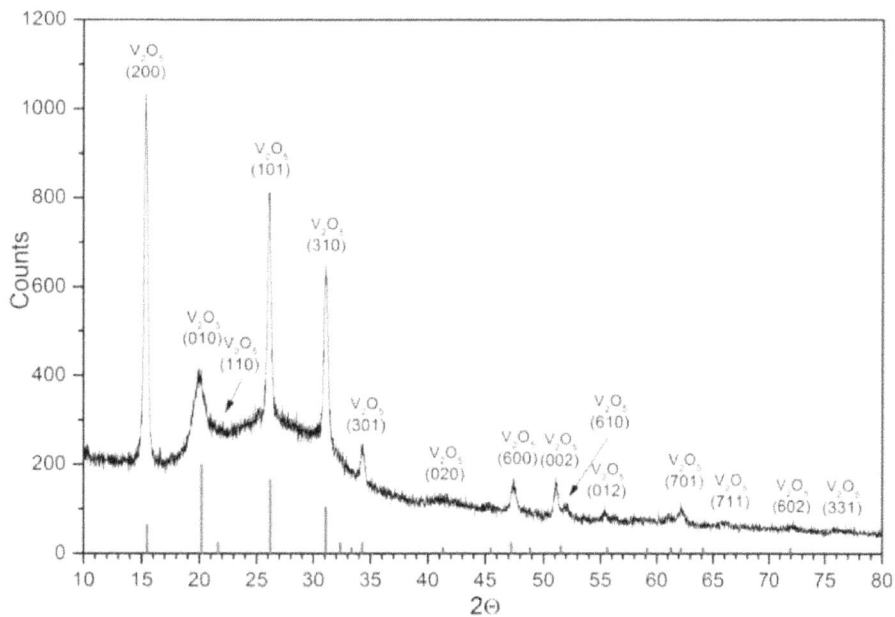

Fig. 4 X-ray diffraction pattern recorded for a Series 3 thin film obtained after 360 min of deposition ($d = 630$ nm); reference lines taken from Ref. [74]

Fig. 5 demonstrates the X-ray diffraction patterns recorded for as-sputtered thin films from Series 1-3. The presented patterns differ of crystalline structure. This can be attributed to the differences in oxygen concentration in the deposition chamber. According to Table 1 the concentrations of oxygen are 4.55%, 9.46% and 26.3% O_2 for Series 1, 2 and 3, respectively. The degree of crystallinity improves with oxygen concentration.

The experimental data were interpreted using XRD patterns distributed by the International Centre for Diffraction DATA – ICDD. As can be seen, the films were poorly crystallized. Using peaks (101), (400), (301) and (221), the V_2O_5 orthorhombic phase (space group Pmmn) was identified. The determined lattice parameters (Table 2) are highly consistent with those taken from literature reports [74-77]

Table 2. V_2O_5 lattice parameters determined in the present work for the Series 3 thin films with different thickness compared with the values reported by other authors [74-77].

Lattice constant	Own work [nm]		Literature data [nm]			
	$d = 430$ nm	$d = 630$ nm	[74]	[75]	[76]	[77]
a	1.150 ± 0.012	1.1448 ± 0.0028	1.148	1.1519	1.1512	1.1375
b	0.441 ± 0.009	0.4359 ± 0.0041	0.436	0.4373	0.4368	0.4318
c	0.357 ± 0.006	0.3549 ± 0.0032	0.355	0.3564	0.3564	0.3519

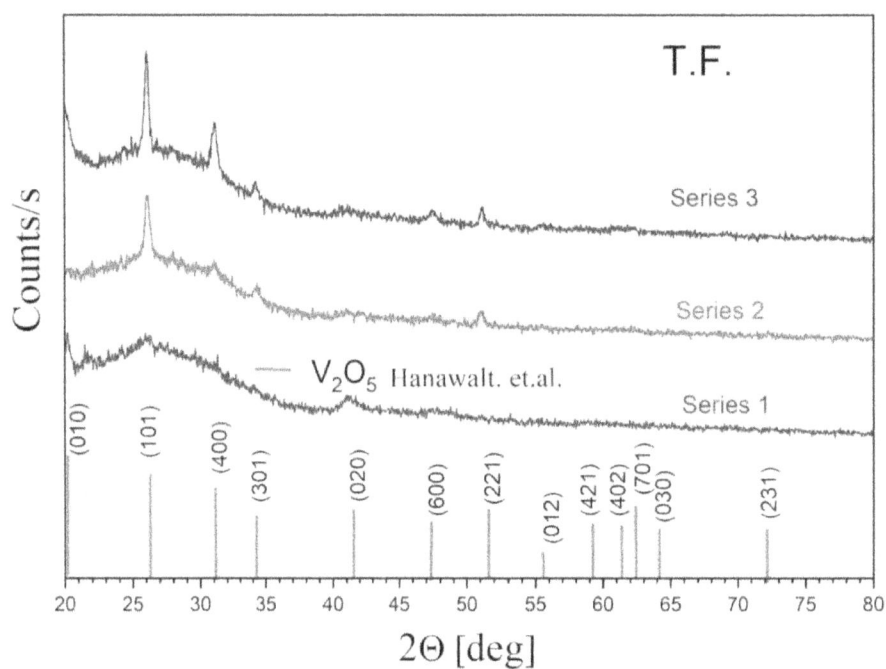

Fig. 5 X-ray diffraction patterns recorded for thin films from Series 1-3. Reference lines taken from Ref. [74].

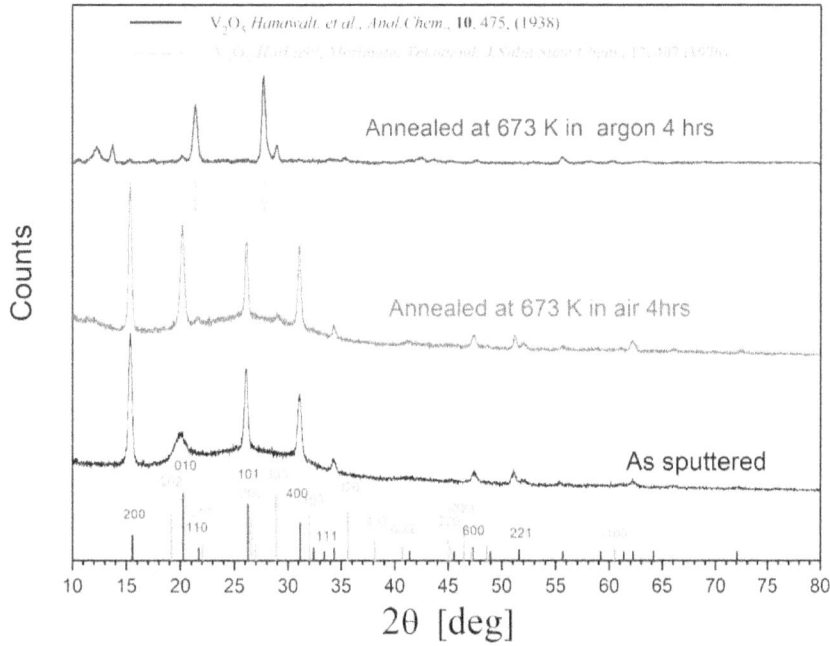

Fig. 6 X-ray diffraction pattern recorded for the Series 3 thin film with $d = 630$ nm, as-sputtered and annealed at 673 K in either air or argon; blue and green reference lines taken from Refs [74] and [78], respectively.

Fig. 6 presents X-ray diffraction patterns recorded for the thicker Series 3 thin film – as-sputtered and annealed for 4 h in either air or argon at 673 K. The sample annealed in air retains the form of vanadium pentoxide. On the other hand, the sample annealed in argon undergoes reduction to the V_3O_5 phase, according to the reaction:

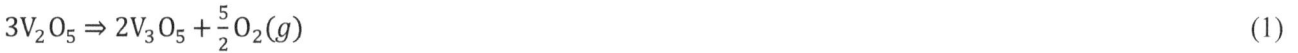

$$3V_2O_5 \Rightarrow 2V_3O_5 + \frac{5}{2}O_2(g) \tag{1}$$

The thermodynamic conditions of the V_2O_5 reduction according to Eq. (1) are described in paper [79].

Crystallite size (d_{XRD}) was calculated from the X-ray broadening of selected peaks, according to Scherrer's method:

$$d_{XRD} = \frac{0.9\lambda}{\Delta(2\theta)\cos\theta} \tag{2}$$

where $\lambda = 0.154056$ nm is the wavelength used for XRD (CuK$_\alpha$), $\Delta(2\theta)$ denotes the broadening of the XRD peak at half of $\Delta(2\theta)$ denotes the broadening of the XRD peak at half of its maximum intensity, and θ represents the Bragg diffraction its maximum intensity, and θ represents the Bragg diffraction angle. The d_{XRD} parameter was determined for four or five most intensive peaks. The determined values are presented in Table 3.

Table 3 Crystallite size (d_{XRD}) determined from Scherrer's equation (2)

Sample	d_{XRD} [nm]
Series 2, as-sputtered	26.2 ± 3.1
Series 3, as-sputtered	24.0 ± 2.4
Series 3, annealed for 4 h in air at 673 K	19.9 ± 1.7
Series 3, annealed for 4 h in Ar at 673 K	21.4 ± 1.4

3.1.1 Series 1

Scanning electron micrographs of the Series 1 thin film and the corresponding EDS spectrum are presented in Fig. 7 a-d and e, respectively. Its grains had a flower-like morphology and a diameter of ca. 250 nm. EDS chemical analysis detected elements originating from the Corning substrate (Si, K, Na and O) and the presence of vanadium and oxygen originating from the film. A detailed analysis of the EDS spectra recorded for different sites on the film surface showed the distribution of vanadium and oxygen to be homogenous.

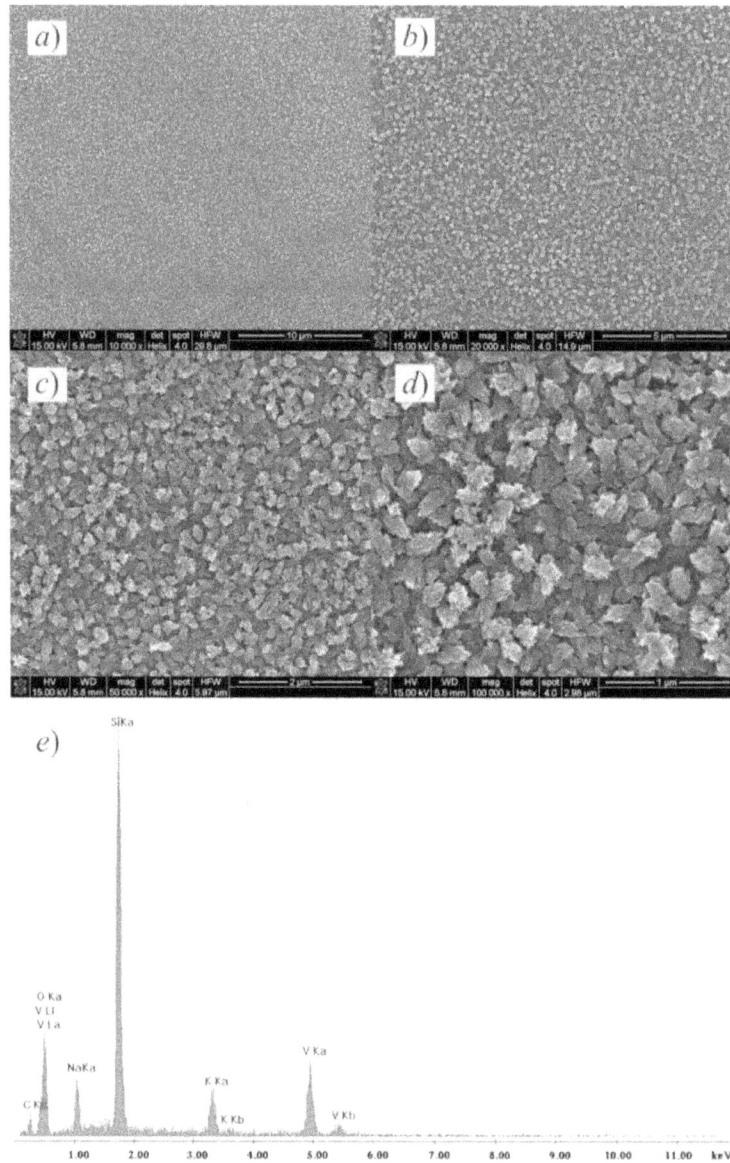

Fig. 7 Scanning electron micrographs of the Series 1 thin film at four different levels of magnification (a-d) and the corresponding EDS spectrum (e).

3.1.2 Series 2

Scanning electron micrographs of the Series 2 thin film and its EDS spectrum are presented in Fig. 8 a-d and e, respectively. As can be seen, the thin film is poly-dispersed, and the grains have a shape of rods with a length of 275 ± 88 nm and a diameter of 51 ± 17 nm.

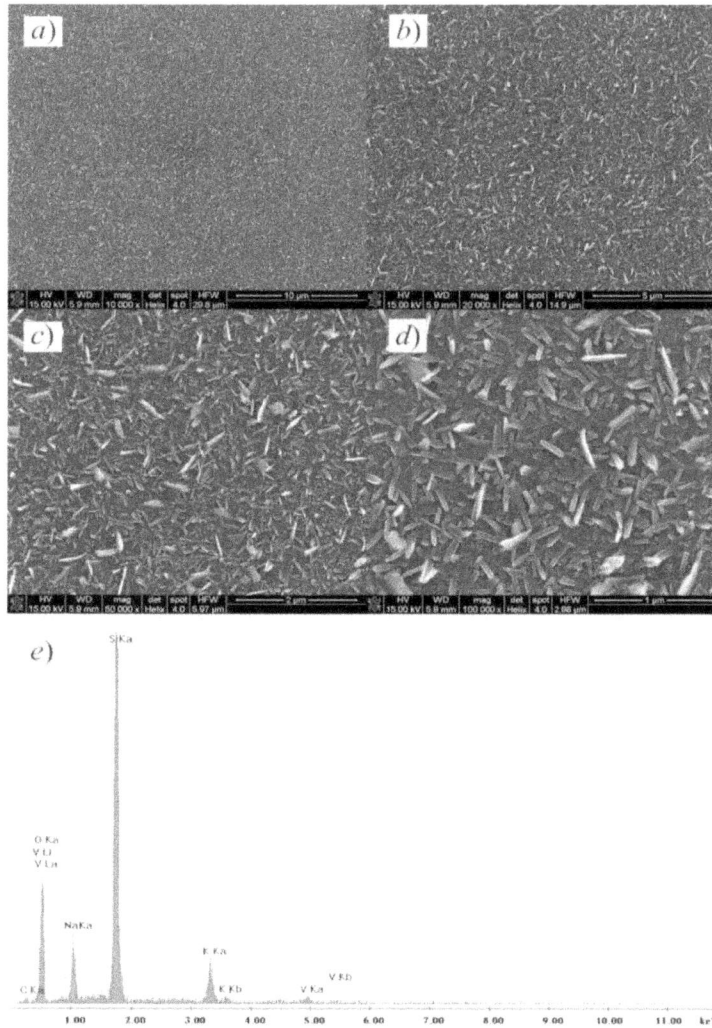

Fig. 8 Scanning electron micrographs of the Series 2 thin film at four different levels of magnification (a-d) and the corresponding EDS spectrum (e).

EDS chemical analysis detected elements originating from the Corning substrate (Si, Na and O) and the presence of vanadium and oxygen originating from the film. As in the case of Series 1, the distribution of vanadium and oxygen was homogenous.

3.1.3 Series 3 (430 nm)

Scanning electron micrographs of the Series 3 thin film and the corresponding EDS spectrum are presented in Fig. 9 a-d and e, respectively. Its grain were spindle-shaped, with a length of 630 ± 35 nm in length and a width of 146 ± 12 nm. Such a microstructure may favourite the postulated preferred crystallographic orientation. EDS chemical analysis detected elements originating from the Corning substrate (Si, K, Na and O) and the presence of vanadium and oxygen originating from the film. The distribution of vanadium and oxygen was again homogenous.

Table 4 Grain size determined from SEM observations

Parameter		Series 1	Series 2	Series 3
a [nm]		N.A.	1.151	1.1449
b [nm]		N.A.	0.443	0.436
c [nm]		N.A.	0.356	0.355
Grain size	length [nm]	232 ± 39	275 ± 88	630
	width [nm]	148 ± 29	51 ± 17	146

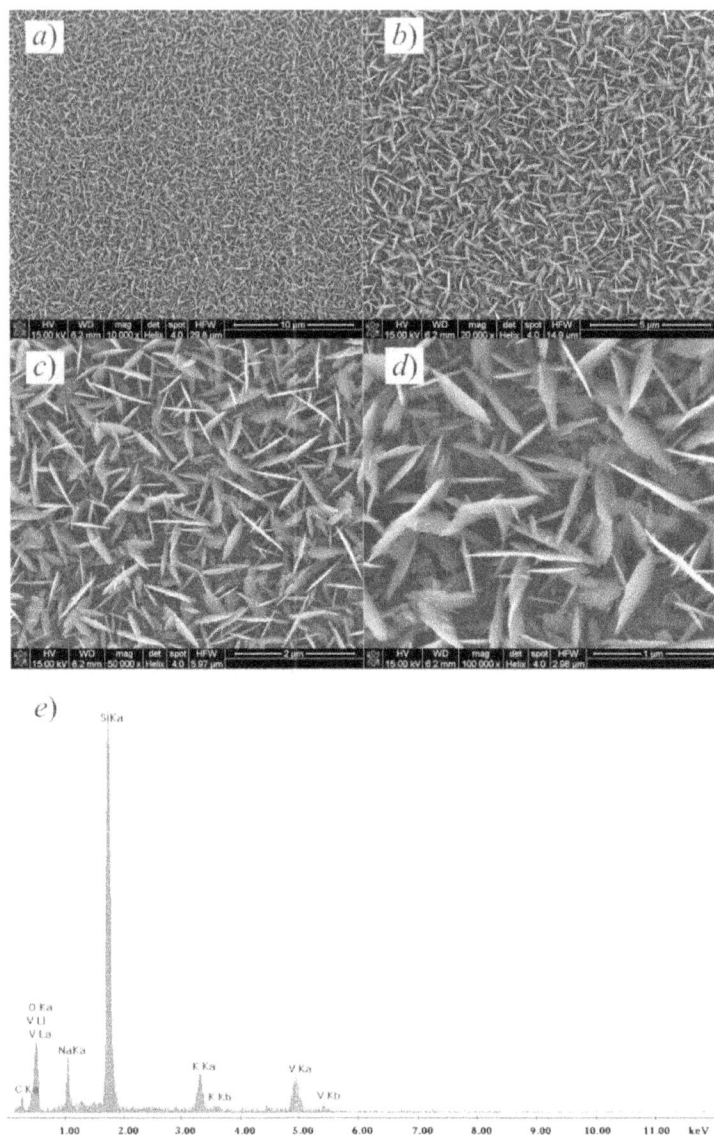

Fig. 9 Scanning electron micrographs of the 430 nm Series 3 thin film at four different levels of magnification (a-d) and the corresponding EDS spectrum (e).

Table 4 summarizes the grain sizes of the thin films (Series 1-3) determined from SEM observations.

The Raman spectra for two films with a thickness of 200 nm and 630 nm are presented in Fig. 10a and Fig. 10b, respectively. From the comparison of the two sets of the data, it can be concluded that thickness does not have a pronounced effect on the Raman spectra recorded for films. The thermal treatment and the resulting crystallization of V_2O_5 are far more significant [80]. The observed results are consistent with the literature data [77-83] (Table 5).

Peaks located at 140, 284, 398, 483, 525, 700 and 990 cm^{-1} can be indexed to the Raman signature of the V_2O_5 crystal [84]. All peaks presented in Fig. 10 can be attributed to the typical vibrations of crystalline V_2O_5 [80-86].

In particular, V-O-V stretching and bending modes are assigned to frequencies from 280 cm^{-1} to 720 cm^{-1} (Fig. 10 a: 141 cm^{-1}, 284 cm^{-1}, 399 cm^{-1}, 484 cm^{-1}, 522 cm^{-1} and 698 cm^{-1}; Fig. 10 b: 140 cm^{-1}, 288 cm^{-1}, 398 cm^{-1}, 483 cm^{-1}, 528 cm^{-1} and 694 cm^{-1}) [4.69, 4.72, 4.73]. The high-frequency peaks located near 1000 cm^{-1} (994 cm^{-1} and 989 cm^{-1} in Figs 10 a and b, respectively) correspond to the terminal oxygen V=O stretching, which results from unshared oxygen atoms [80, 87, 88].

58

Table 5 Raman modes of vanadium pentoxide thin films and nanorods [cm⁻¹]

Table 5 Raman modes of vanadium pentoxide thin films and nanorods [cm^{-1}]

Present work		Lee et al. [77, 80]		Su et al. [81]	Su et al. [82]	Schrecken-bach et al. [83]	Dong et al. [84]
		RF sputtering		CVD	DC	Electrochem.	VO_x/TiO_2
$d = 200$ nm	$d = 600$ nm	Amorphous	crystalline	rod	magnetron	deposition	catalysts
141.4	139.6	-	142	145	146	143	282
283.9	287.9	-	283	284	286	-	404
398.6	398.1		405	406	404	405	-
483.9	482.6		487	484	-	-	524
521.8	528.3	520	530	531	520	526	699
697.8	693.6	650	706	704	706	-	993
994.5	988.9	932	1000	996	992	983	
		1027					

a

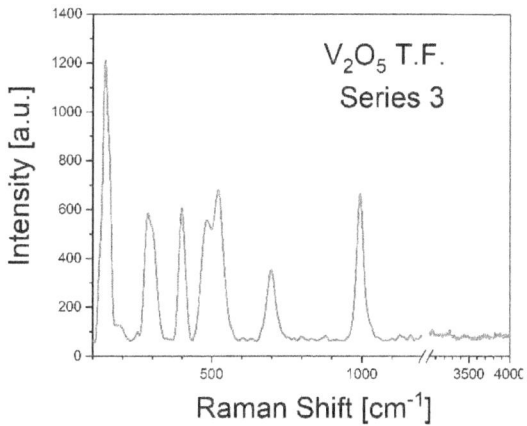

b

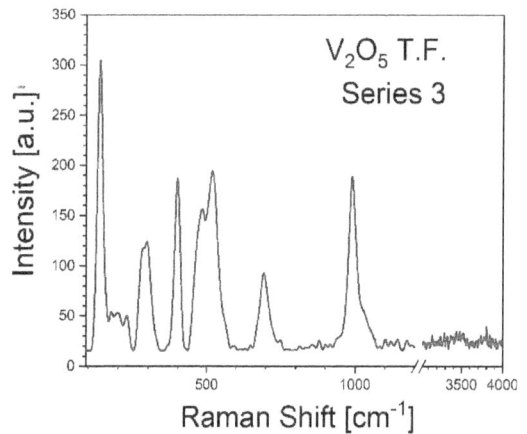

Fig. 10 Raman spectra recorded for V_2O_5 thin films: a) $d = 200$ nm, b) $d = 630$ nm

X-ray reflectometry (XRR) was used as a complementary method for the evaluation of film thickness and density [61]. The parameters determined from Fig. 11 are shown in Table 6.

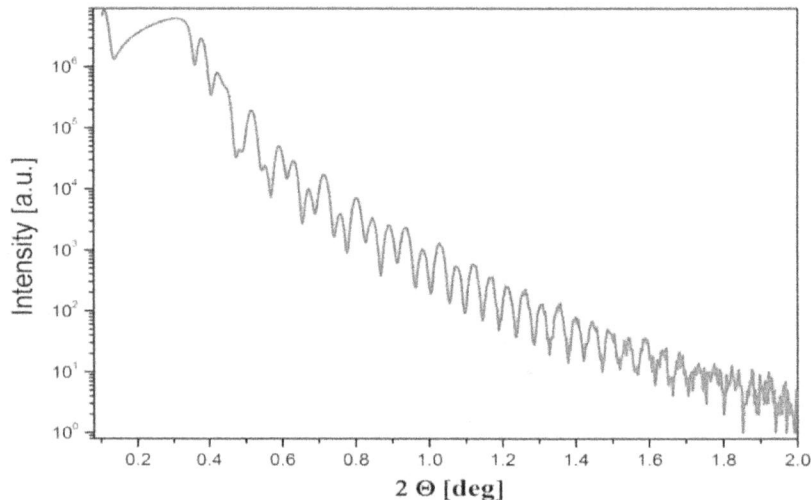

Fig. 11 XRR curve recorded for the as-sputtered thin film

Table 6. Parameters of the as-sputtered vanadium oxide thin film

1	2	3	4
Phase	Thickness [nm]	Roughness [-]	Density [g/cm^3]
Substrate (fused silica)	-	1.5	2.66
Vanadium oxide	400	12	3.01

The determined density values of the substrate (2.66 g/cm^3) and thin film (3.01 g/cm^3) are similar to the literature data [89] for fused silica (2.635 – 2.660 g/cm^3) and vanadium pentoxide (3.357 g/cm^3).

4. CONCLUSIONS

Several vanadium oxide thin film deposition techniques were reviewed. The technique selected for the deposition of vanadium pentoxide thin films in the presented study was reactive radio frequency sputtering. The structure and morphology of four series of V$_2$O$_5$ thin films prepared under different conditions were described.

REFERENCES

1. S. Beke, *A review of the growth* V$_2$O$_5$ *films from 1885 to 2010*, Thin Solid Films 519 (2011) 1761-1771 and references [10-125] ibidem.
2. Nag, R.F. Haglund Jr, *Synthesis of vanadium dioxide thin films and nanoparticles*, J.Phys.: Condens. Mater. 20 (2008) 1-14.
3. D. M. Mattox, *Handbook of Physical vapor deposition (PVD) Processing 2nd* 2010, Edition, Elsevier Inc.
4. J-G. Zhang, P. Liu, J.A. Turner, C.E. Tracy, D.K. Benson, *Highly stable vanadium oxide cathodes prepared by plasma-enhanced chemical vapor deposition*, J. Electrochem. Soc. 145 (1998) 1889-1892
5. D. Barreca, L. Armelao, F. Caccavale, V. Di Noto, A. Gregori, G. A. Rizzi, E. Tondello, *Highly oriented V$_2$O$_5$ nanocrystalline thin films by plasma-enhanced chemical vapor deposition,* Chem. Mater.12 (2000) 98-103.
6. J. Musschoot,, D. Deduytsche, H. Poelman, J. Haemers, R. L. Van Meirhaeghe, Van den Berghe, C. Detavernier, *Comparison of Thermal and Plasma-Enhanced ALD/CVD of Vanadium Pentoxide,* J.Electrochem.Soc. 156 (2009) 122-126.
7. M. George, *Atomic layer deposition: An overview*, Chem.Rev. 110 (2010) 111-131.

8. S. Boukhalfa, K. Evanoff, G. Yushin, *Atomic layer deposition of vanadium oxide on carbon nanotubes for high-power supecapacitor electrodes*, Energy Environ. Sci, 5 (2012) 6872-6879

9. M.B. Sreedhara, J. Ghatak, B. Bharath, C.N.R. Rao, *Atomic laser deposition of ultrathin crystalline epitaxial films of V_2O_5*, ACS Applied Mater. Interfaces 9 (2017) 3178-3185.

10. L.A. Ryabova, I.A. Serbinov, A.S. Danevski, *Preparation and properties of vanadium oxide films.* J. Electrochem.Soc. 119 (1972) 427-429.

11. T. Maruyama, Y. Ikuta, *Vanadium dioxide thin films prepared by chemical vapour deposition frm vanadium(III) acetyloacetonate*, J. Mater.Sci. 28 (1993) 5073-5078.

12. C. Picirillo, R. Bininions, I.P. Parkin, Nb-*doped VO_2 thin film prepared by aerosolo-assisted chemical vapour deposition*, Eur.J. Inorg. Chem 25 (2007) 4050-4055.

13. H. Takei, S. Koide, *Growth and electrical properties of vanadium-oxide single crystals by oxychloride decomposition method*, J.Phys.Soc.Japan 21 (1966) 1010.

14. T.D. Manning, I.P. Parkin, C. Blackman, U. Qureshi, *APCVD of thermochromic vanadium dioxide thin films-solid solutions $V_{2-x}M_xO_2$ (M = Mo, Nb) or composites VO_2: SnO_2*, J.Mater. Chem. 15 (2005) 4560-4566.

15. M. Losurdo, D. Barreca, G. Bruno, E. Tondello, *Spectroscopic ellipsometry investigation of V_2O_5 nanocrystalline thin films,* Thin Solid Films 384 (2001) 58-64.

16. G. Rampelberg, D. Deduytsche, B. De Schutter, P. A. Premkumar, M. Toeller, M. Schaekers, K. Martens, I. Radu, Ch. Detavernier, *Crystallization and semiconductor-metal switching behavior of thin VO_2 layers grown by atomic layer deposition,* Thin Solid Films 550 (2014) 59-64.

17. J. Musschoot, D. Deduytsche, H. Poelman, J. Haemers, R. L. Van Meirhaeghe, S. Van den Berghe, C. Detavernier, *Comparison of thermal and plasma-enhanced ALD/CVD of vanadium pentoxide*, J. Electrochem. Soc. 156 (2009) 122-126.

18. X.Y. Chen, E. Pomerantseva, P. Banerjee, K. Gregorczyk, R. Ghodssi, G. Rubloff, *Ozone-based atomic layer deposition of crystalline V_2O_5 films for high performance electrochemical energy storage,* Chem. Mater. 24 (2012) 1255-1261.

19. I. I. Kazadojev, S. O'Brien, L. P. Ryan, M. Modreanu, P. Osiceanu, S. Somacescu, D. Vernardou, M. E Pemble and I. M Povey, *Growth of V_2O_5 films for battery applications by pulsed chemical vapor deposition*, ESC Trans. 85 (2018) 83-94.

20. M. Abbasi, S.M. Rozati, R. Irani, S. Beke, *Synthesis and gas sensing behavior of nanostructured V_2O_5 thin films prepared by spray pyrolysis,* Materials Science in Semiconductor Processing 29 (2015) 132-138.

21. M. Basha, L. Akilasundari, *Effect of Substrate Temperature on Structural, Optical and Surface Morphological Properties of Spray Deposited V_2O_5, Thin Film*, Research Journal of Pharmaceutical, Biological and Chemical Sciences 4 (2013) 169-175.

22. C. Ban, N.A. Chernova, M.S. Whittingham, *Electrospun nano-vanadium pentoxide cathode*, Electrochem. Comm. 11 (2009) 522-525.

23. R. Ostermann, D. Li, Y. Yin, J.T. Mcann, Y. Xia, *V_2O_5 nanorods on TiO_2 nanofibres: A new class of hierarchical nanostructures enabled by elecrtrospining and calcinations* Nanoletters 6 (2006) 1297-1302.

24. L. Mai, X. Xu, L. Xu, Ch. Han, Y. Luo, *Vanadium oxide nanowires for Li-ion batteries*, J. Mater.Res. 26 (2011) 2175.2185.

25. C. Diaz, G. Barrera, M. Segovia, M.L. Valenzuela, M. Osiak and C. O'Dwyer, *Crystallizing vanadium pentoxide nanostructures in the solid-state using modified block copolymer and chitosan complexes*, J. Nanomater. 2015 (2015) Article ID 105157, 13.

26. A. Akl, *Effect of solution molarity on the characteristics of vanadium pentoxide thin film*, Appl. Surf. Sci. 252 (2006) 8745-8750.

27. A. Lahiri, F. Enders, *Electrodeposition of nanostructured materials from aqueous, organic and ionic liquid electrolytes for Li-ion and Na-ion batteries: A comparative review*, J. Electrochem. Soc. 164 (2017) D597-D612.

28. Y.-R. Lu, T.-Z. Wu, Ch.-L. Chen, D-H. Wei, J-L. Chen, W-Ch. Chou, Ch-L Dong, *Mechanism of electrochemical deposition and coloration of electrochromic V_2O_5 nano thin film: as in situ X-ray spectroscopy study*, Nanoscale Research Letters 10 (2015) 387.

29. M. Rasoulis, D. Varnardou, *Electrodeposition of vanadium oxides at room temperature as cathodes in lithium-ion batteries*, Coatings 7 (2017). 8.doi:10.3390/coatings7070100.

30. O.M. Osmolovskaya, I.V. Murin, V.M. Smirnov, M.G. Osmolovsky, *Synthesis of vanadium dioxide thin films and nanpowders: a brief review*, Rev.Adv.Mater.Sci. 36 (2014) 70-74.

31. J. Wu. W. Huang, Q. Shi, J. Cai, D. Zhao, Y. Zang, J. Yan, *Effect of annealing temperature on thermochromic properties of vanadium dioxide thin film deposited by organic sol-gel method,* Appl.Surf.Sci. 268 (2013) 556-560.

32. B.G. Chae, H.T. Kim, *Effects of W doping on the metal-insulator transition in vanadium dioxide film*, Physica B 405 (2010) 663-667.

33. J.-H. Cho, Y-J. Byun, J-H. Kim, Y-J. Lee, Y-H. Jeong, M.-P. Chun, J-H. Paik, T.H. Sung, *Thermochromic characteristics of WO_3.doped vanadium dioxide thin films prepared by sol-gel method*, Ceram.Intern. 38 (2012) 589-593.

34. M. Pan, H. Zhong, S. Wang, J. Liu, Z. Li, X. Chen, W. Lu, *Properties of VO_2 thin film prepared with precursor VO (acac)₂*, J. Cryst. Growth 265 (2004) 121-126.

35. T.J. Hanlon, J.A. Coath, M.A. Richardson, *Molybdenum-doped vanadium dioxide coatings on glass produced by the aqueous sol-gel method*, Thin Solid Films 436 (2003) 269-272.

36. Y, Ningvi, Jinhua, L. Chenglu, *Valence reduction process from sol-gel V_2O_5 to VO_2 thin films*, Appl.Surf.Sci. 191 (2002) 176-180.

37. Z.S. El Mandouh, M.S. Selim, *Physical properties of vanadium pentoxide sol gel films,* Thin Solid Films 371 (2000) 259-263.

38. M.C. Rao, K. R. Rao, *Thermal Evaporated V_2O_5 Thin Films*: Thermodynamic Properties, International Journal of Chem.Tech Research 6 (2014) 3931-3934.

39. M. Singh, R.K. Sharma, G, B. Reddy, *Growth of α-V_2O_5 nanostructured thin films as a function of deposition process,* AIP Conference Proc. 1731 (2016) 050041 p.1.3.

40. G. Wu, K. Du, Ch. Xia, X. Kun, J. Shen, B. Zhou, J. Wang, *Optical absorption edge evolution of vanadium pentoxide films during lithium intercalation*, Thin Solid Films 485 (2005) 284-289.

41. A. Kumar, P. Singh, N. Kulkarni, D. Kaur, *Structural and optical studies of nanocrystalline V_2O_5 thin films,* Thin Solid Films 516 (2008) 912-918.

42. N.M. Abd-Alghafour, N.M. Ahmed, Z. Hassan, S.M Mohammad, *Influence of solution deposition on properties of V_2O_5 thin films deposited by spray pyrolysis technique*, AIP Conference Proc. 1756 (2019) 10.1063/1.4958791.

43. C.V. Ramana, O.M. Hussain, S. Uthanna, B.S. Naidu, *Influence of oxygen partial pressure on the properties of electron beam evaporated vanadium pentoxide thin films*, Opt.Mater. 10 (1998) 101-107.

44. A. Laurenco, A. Gorenstein, S. Passerini, W.H. Smyrl, M.C.A. Fantini, M.H. Tabacniks, *Radio-frequency reactively sputtered VO_x thin films deposited at different oxygen flows*, J. Electrochem. Soc. 45 (1998) 706-711.

45. J.Huotari, R. Bjorklund, J. Lappalainen, A. Lloyd Spetz, *Pulsed Laser Deposited Nanostructured Vanadium Oxide Thin Films Characterized as Ammonia Sensors*, Sensors and Actuators B 217 (2015) 22-29.

46. S. Beke, S. Giorgio, L. Körösi, L. Nanai, W. Marine, *Structural and optical properties of pulsed laser deposition V_2O_5 thin films,* Thin Solid Films 516 (2008) 4659-4664.

47. M. Bore, F. Qian, F. Nagabushnam, R.K. Singh, *Pulsed laser deposition of oriented VO_2 thin films on R-cut sapphire substrates*, Appl.Phys.Lett. 63 (1993) 3288-3290.

48. J.Y. Suh, R. Lopez, F.C. Feldman, R.F. Jr. Haglund, *Semiconductor to metal phase transition in the nucleation and growth of* VO_2 *nanoparticles and thin film*, J. Appl.Phys. 96 (2004) 1209-1213.

49. E. Prociow, M. Zielinski, K. Sieradzka, J. Domaradzki, D. Kaczmarek, Electrical and optical study of transparent V-based semiconductor prepared by magnetron sputtering under different conditions, Radioengineering 20 (2011) 204-208.

50. M.S.B. de Castro, C.L. Ferreira, R.R. Avillez, *Vanadium oxide thin films produced by magnetron sputtering from a* V_2O_5 *target at room temperature*, Infrared Phys. Technolog. 60 (2013) 103-107.

51. P. D. Raj, S. Gupta, M. Sridharan, *Studies on nanostructured* V_2O_5 *deposited by reactive dc magnetron sputtering, Physics and semiconductor devices, environmental science and engineering,* V.K. Jain & A. Verma, Eds, Springer Int.Publ. Switzerland 2014, 573-576.

52. D. Palai, C.M. Esher, D. Porwal, A. Dey, *Reversible phase transition in vanadium films sputtered on metal substrates*, Phil. Mag.Lett.96 (2016) 440-446.

53. H.M.R. Giannetta, C. Calaza, L. Fraigi, L. Foneca, *Vanadium oxide thin films obtained by thermal annealing of layers deposited by rf magnetron sputtering at room temperature, in: Modern technologies for creating the thin-film systems and coatings*, N.N. Nikitenov, Ed. ch.9 (2017).

54. A. Gies, B. Pecquenard, A. Benayad, H. Martinez, D. Gonbeau, H. Fuess, A. Levasseur, *Effect of total gas and oxygen pressure during deposition on the properties sputtered* V_2O_5 *thin films.* Solid State Ionics 176 (2005) 1627-1634.

55. D. Rachel Malini, C. Sanjeeviraja, *Effect of RF power on electrochromic* V-Ce *mixed oxide thin films*, Electrochim. Acta 104 (2013) 162-169.

56. M. Jiang, Y. Li, S. Li, H. Zhou, X. Cao, S. Bao, Y. Gao, H. Luo, P. Jin, *Room Temperature Optical Constants and Band Gap Evolution of Phase Pure M1.* VO_2 *Thin Films Deposited at Different Oxygen Partial Pressures by Reactive Magnetron Sputtering*, J. Nanomater. (2014) 1-6.

57. L. Zhang, J. Tu, H. Feng, J. Cui, *Study of mixed vanadium oxide thin film deposited by RF magnetron sputtering and its application*, Phys. Procedia 18 (2011) 73-76.

58. D. Acousta, A. Perez, C. Magana, F. Hernandez, V_2O_5 *thin film deposition by rf magnetron sputtering. The influence of oxygen content in physical properties*, J.Mater.Sci. Eng. A6 (2016) 81-87.

59. L.J. Meng, R.A. Silva, H.N. Cui, V. Teixeira, M.P. Dos Santos, Z. Xu, *Optical and structural properties of vanadium pentoxide films prepared by d.c. reactive magnetron sputtering*, Thin Solid Films 515 (2006) 195-200.

60. *G. Micocci, A. Serra, A. Tepore, and S. Capone, R. Rella and P. Siciliano, Properties of vanadium oxide thin films for ethanol sensor*, Citation: Journal of Vacuum Science & Technology A 15, (1997) 34.

61. K. Schneider, K. Zakrzewska, Z. Tarnawski, K. Drogowska, N.T.H. Kim-Ngan, VO_x *thin films deposited by reactive r.f. sputtering,* Ceramics (2013).

62. M. Benmoussa, E. Ibnouelghazi, A. Bennouna, E.L. Ameziane, *Structural, electrical and optical properties of sputtered vanadium pentoxide thin films,* Thin Solid Films 265 (1995) 22-28.

63. R.M. Öksüzoğlu, P.Bilgic, M.Yildrim, O. Deniz,, *Influence of post-annealing on electrical, structural and optical properties of vanadium oxide thin films*, Optics & Technology 48 (2013) 102-109.

64. M. Singh, R.K.Sharma, G.B. Reddy, *Growth of α-* V_2O_5 *nanostructured thin films as a function of deposition process*, AIP Conf. Proc. 1731 (2016) 050041.

65. Y.J. Park, K.S. Ryu, M.G. Kang, M Gu Kang, S. Ho Chang, *Electrochemical properties of vanadium oxide thin film deposited by R.F. sputtering*, Solid State Ionics 154-155 (2002) 229-235.

66. E. Cazanelli, G. Mariotto, S. Passerini, W.H. Smyrl, A. Gorenstein, *Raman and XPS characterization of vanadium oxide thin films deposited by reactive RF sputtering*, ChemEng. Mater. Sci. 56 (1999) 249-258.

67. T.J. Hanlon, R.E. Walker, J.A. Coath, M.A, Richardson, *Comparison between vanadium dioxide coatings on glass produced by sputtering, alkoxide and aqueous sol-gel methods*, Thin Solid Films 495 (2002) 234-237.

68. J.D. Hanawalt, H.W. Rinn, L.K. Fravel, *Chemical analysis by X-ray diffraction*, Anal. Chem. 10 (1938) 475-512.

69. K. Schneider, *Structural and optical properties of* VO_x *thin films*, Archives of Metallurgy and Materials / Polish Academy of Sciences. Committee of Metallurgy. Institute of Metallurgy and Materials Science 60 (2A) (2015) 957-960.

70. K. Schneider, M. Lubecka, A. Czapla, VO_x *thin films for gas sensor applications*, Procedia Engineering 120 (2015) 1153-1157.

71. K. Schneider, M. Lubecka, A. Czapla, V_2O_5 *thin films for gas sensor applications*, Sensors and Actuators B: Chemical 236 (2016) 970-977.

72. K. Schneider, *Defect structure of* V_2O_5 *thin film gas sensor*, Proc. of SPIE 10161 (2017) 1016109.1-10161.9.

73. K. Schneider, M. Dzibaniuk, J. Wyrwa, *Impedance spectroscopy of vanadium pentoxide thin films*, J. Electronic Materials (2019) 4177. DOI:10.1007/s11664-019-07166-x

74. K. Schneider, W. Maziarz, *Vanadium pentoxide thin films: synthesis, characterization and nitrogen oxide sensing properties*, Sensors 18 (2018) 4177.

75. R. Enjalbert, J. Galy, *A refinement of the structure of* V_2O_5, Acta Crystallogr. Sect C Cryst. Struct. Commun. 42 (1986) 1467-1469.

76. A. Chakarbarti, K. Hermann, R. Druzinic, M. Witko, F. Wagner, M. Petersen, *Geometric and electronic structure of vanadium pentoxide: a density functional bulk and surface study*, Phys. Rev. B 16 (1999) 10583-10590.

77. S.H. Lee, H.M. Cheong, M.J. Seong, P. Liu, C.E. Tracy, A. Mascarenhas, *Microstructure study of amorphous vanadium oxide thin films using Raman spectroscopy*, J. Appl. Phys. 92 (2002) 1893-1897.

78. S.H. Lee, H.M. Cheong, M. J. Seong, P. Liu, C. E. Tracy, A. Mascarenhas, J. R. Pitts, S. K. Deb, *Raman spectroscopic studies of amorphous vanadium oxide thin films*, Solid State Ionics 165 (2003) 111-116.

79. K Schneider, *Vanadium oxides –properties and applications. Part II*, this issue

80. Q. Su, C.K. Huang, Y. Weng, Y.C. Fan, B.A. Lu, W. Lan, Y.Y. Wang, X.Q. Liu, *Formation of vanadium oxides with various morphologies by chemical vapor deposition*, J. Alloys Compd. 475 (2009) 518-523.

81. Q. Su, Q. Liu, M.L. Ma, Y.P. Guo, Y.Y. Wang, *Raman spectroscopy characterization of the microstructure of* V_2O_5 *films*, J. Solid State Electrochem. 12 (2008) 919-923.

82. J.P. Schreckenbach, K. Witke, D. Butte, G. Marx, *Characterization of thin metastable vanadium oxide films by Raman spectroscopy*, Fresenius J. Appl. Chem. 363 (1999) 211-214.

83. L. Dong, C. Sun, C. Tang, B. Zhang, J. Zhu, B. Liu, F. Gao, Y. Hu, L. Dong, Y. Chen, *Investigation of surface* VO_x *species and their contributions to activities of* $VO_x/Ti_{0.5}Sn_{0.5}O_2$ *catalysts towards selective catalytic reduction of* NO by NH_3, Appl. Catal. A: General 431-432 (2012) 126-136.

84. J. Due-Hansen, S. Boghosian, A. Kustov, P. Fristrup, G. Tsilomelekis, K. Stahl, C.H. Christensen, R. Fahrmann, *Vanadium-based SCR catalysts supported on tungstated and sulfated zirconia: influence of doping with potassium*, J. Catal. 251 (2007) 459-473.

85. A. Gomstein, A. Khelfa, J.P. Guesdon, C. Julien, *Effect of crystallinity of* V_6O_{13} *films on the electrochemical behaviour of lithium microbatteries*, Mat.Res.Symp.Proc. 369 (1995) 649-655.

86. C. Julien, G.A. Nazari, O. Bergström, *Raman scattering of microcrystalline* V_6O_{13}, Phys.Status Solidi (b) 201 (1997) 319-326.

87. R.J.H. Clark in: *The chemistry of titanium and vanadium*, Elsevier, New York 1968.

88. H. Horiuchi, N. Morimoto, M. Tokonami, *Crystal structures of* V_nO_{2n-1} ($2 \le n \le 7$), J. Solid State Chem. 17 (1986) 407-424.

89. *Handbook of chemistry and physics*, CRC Press Inc., 60-th Edition (1980), Boca Raton, Florida 33431.

V. Electronic structure

Abstract

The electronic structure of the three main vanadium oxides – V_2O_3, VO_2 and V_2O_5 – is reviewed. The optical properties of vanadium pentoxide thin films were determined. It was found that a direct allowed (DA) transition is the most probable one in the studied vanadium pentoxide thin films.

Key words: Vanadium oxides, optical properties, thin films, energy band gap

1. INTRODUCTION

Correlated electrons in vanadium oxides are responsible for their unique structural, electrical, optical and magnetic properties. Their electronic band structures are affected by crystallographic structure, crystal field splitting and hybridization between O 2p and V 3d bands. There have been many experimental and theoretical studies of the band structure of the main vanadium oxides – V_2O_3, VO_2 and V_2O_5. The first experimental studies on band structure were based on optical spectroscopy that utilized absorption and reflection of light. Ceramic materials, single crystals and thin films were the subjects of these studies. Lately, new methods such as photoemission spectroscopy (PES), X-ray absorption or emission spectroscopy (XAS, XES) [1], X-ray reflectivity (XRR), X-ray fluorescence (XRF) [2], photoluminescence (PL) [3], Raman scattering, and scanning tunnelling microscopy (STM) [4] have been used.

Theoretical calculations used several quantum mechanics models such as the Hartree-Fock self-consistent field method based on one-electron approximation, the Hubbard-Mott model [5] introducing the effects of electron correlations on the Hamiltonian, Peierls mechanism [6,7] involving electron-phonon interactions, or the density function theory (DFT) [8].

One of the most important parameters with regard to the properties of materials is the band-gap energy (E_g). Generally, the E_g of a semiconductor or an insulator has been found to decrease with increasing temperature. The variation of the fundamental E_g with temperature is very important for both basic science and technological applications.

2. ELECTRONIC STRUCTURE

The theoretical basis of optical properties results from Maxwell's equations. From optical spectra, the complex dielectric function $\varepsilon(\omega)$ is derived [9]:

$$\varepsilon(\omega) = \varepsilon_1 - i\varepsilon_2 \tag{1}$$

where ω is angular frequency of light ($\omega = 2\pi\nu$), ε_1 and ε_2 represent real and complex parts of ε, and i is the imaginary unit.

$$\varepsilon_1 = n^2 - \kappa^2 \tag{2}$$

and:

$$\varepsilon_2 = 2n\kappa \tag{3}$$

Measurement of light absorption is one of the most important techniques used to determine the optical properties of solids. In absorption measurements, the intensity of light ($I(d)$) after it has travelled through a certain thickness of a material is compared with the incident intensity (I_o), thereby defining the absorption coefficient (α):

$$I(d) = I_o \exp[-\alpha(\omega)d] \hspace{8cm} (4)$$

The dependence of the absorption coefficient on frequency is shown in Fig. 1. Since $I(d)$ depends on the square of the field variables, it immediately follows that

$$\alpha(\omega) = 2\frac{\omega\,\kappa}{c} = 4\,\frac{\pi\kappa}{\lambda} \hspace{7cm} (5)$$

where the factor of 2 results from the definition of $\alpha(\omega)$ in terms of light intensity, which is proportional to the square of the optical fields. This expression entails that the absorption coefficient is proportional to $\sim\kappa(\omega)$, the imaginary part of the complex index of refraction (extinction coefficient), so that κ is usually associated with power loss.

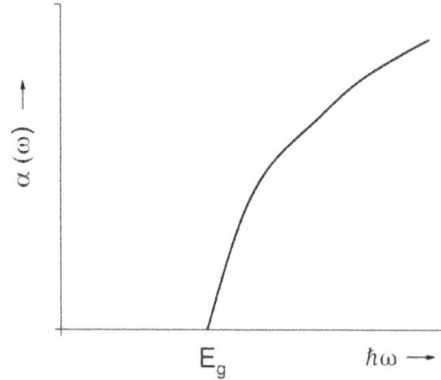

Fig. 1 Frequency dependence of the absorption coefficient (α) near the threshold for interband transition

It can be concluded from Eqs (2), (3) and (5) that either set – $(\varepsilon_1, \varepsilon_2)$ or (n, α) – represents the wavelength-dependent constants characterizing the optical properties of a studied solid. Many efforts have been devoted to the determination of the refractive index (n), but there is so far no universal approach. Several methods of calculating the refractive index of vanadium oxide [10-17] and other materials [18-22] have been proposed. Various assumptions were used in these methods, and the results are often subjective/debatable and ambiguous.

2.1 Optical properties V_2O_5 – literature survey

The electronic structure of vanadium pentoxide has been the subject of intensive studies [23-33]. Various theoretical calculations of V_2O_5 band structure include both semi- empirical and *ab initio* techniques.

Lambrecht et al. [23, 24] presented a calculation of the energy band structure using a tight-binding model in which the oxygen p-bands and vanadium d-bands were decoupled. They applied a perturbation approximation in order to obtain an effective Hamiltonian for the valence and conduction bands separately. The theoretically determined dispersion of the energy bands was verified by applying electrical transport properties. The valence band density of states was compared with XPS (X-ray photoelectron spectroscopy) and SXS (soft X-ray spectroscopy) data. The valence-to-conduction band transitions were compared with optical and electron energy loss (ELS) data. A satisfactory agreement between theoretical and experimental data was found [24].

Kempf et al. [25] reported on pseudopotential periodic Hartree-Fock calculations on a V_2O_5 crystal. The determined V-O bond lengths and stretching force constants were found to be in good agreement with experimental data. The estimated band structure and density of states remain in contrast with tight-binding calculations. There is no gap between the conductor and valence bands. According to the authors, vanadium pentoxide is partially ionic.

D. W. Bullett [26] determined the electronic structure of vanadium pentoxide using direct and non-empirical atomic orbital techniques. He postulated an indirect semiconducting energy gap of 2.6 eV.

V. Eyert and K.-H. Höck [27] computed the band structure of bulk vanadium pentoxide using the density functional theory (DFT) and local-density approximations (LDA). Its electronic properties were modified via strong hybridization between O 2p and crystal-field-split V3d orbitals. A strong deviation of VO_6 octahedra from the cubic coordination led to a narrow split-off conduction band.

The electronic structure of vanadium pentoxide is strongly connected with its anisotropy, which in turn is associated with its crystal structure. The atoms form double chains within planes that are separated by a van der Waals gap (see chapter 2.3.4).

Kenny et al. [28] studied the optical absorption coefficients of V_2O_5 single crystals using incident polarized light with wavelengths in the range of 0.47-1.8 μm and unpolarized light with wavelengths from 1.5 to 7.5 μm. Fundamental absorption was observed at incident photon energies of 2.15, 2.22 and 2.17 eV for $E\|a$, $E\|b$, $E\|c$, respectively. Some evidence for a direct forbidden transition mechanism with band gaps of 2.36 and 2.34 eV was observed for $E\|a$ and $E\|c$, respectively. The most notable property of V_2O_5 is its ability to produce monolayers (or materials only several layers thick) [29]. Vanadium pentoxide is the second material known to exhibit such a property. The first one was graphite forming a single monolayer, known as graphene. Chakrabarti et al. [30] determined the V_2O_5 mono-layer band structure using *ab initio* density-functional theory (DFT). The obtained results are in excellent agreement with experimental crystallographic data as well as with other experimentally determined surface properties [31, 32].

T.M. Tolhurst et al. [33] studied a double-layered polymorph of V_2O_5 (named ε'-V_2O_5) using soft X-ray spectroscopic measurements and density functional theory calculations. This polymorph has increased interlayer separation, which leads to a dramatic increase in the band gap.

Table 1 Band gap (E_g) of V_2O_5 and its temperature dependence – dE_g/dT

E_g [eV]	dE_g/dT [eV/K]	Orientation	References
2.17			[28], [34]
2.34		$E\|c$	[35]
2.25			[36]
2.49 ($T{\rightarrow}0$)	$-6.1\cdot10^{-4}$		[37]
2.19			[34]
2.363		$E\|b$	[35]
2.22			[28]
2.23	7.3	$E\|a$	[36]
2.15			[28]
2.54 ($T{\rightarrow}0$)	$-6.1\cdot10^{-4}$	$E{\perp}c$	[37]

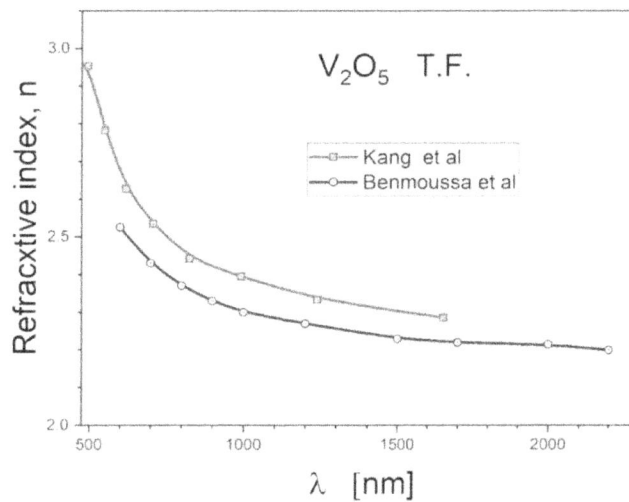

Fig. 2 Refractive index (n) determined for V_2O_5 thin films [38, 39]

Fig. 2 shows the refractive index for vanadium pentoxide thin films as a function of wavelength [38, 39]. Parameter n decreases with wavelength. This dependence may be verified using the theoretical equation proposed by Cauchy [40]:

$$n(\lambda) = A + \frac{B}{\lambda^2} \tag{6}$$

where A and B are independent of λ. According to Fig. 2, the experimental points are consistent with the theory postulated by Cauchy.

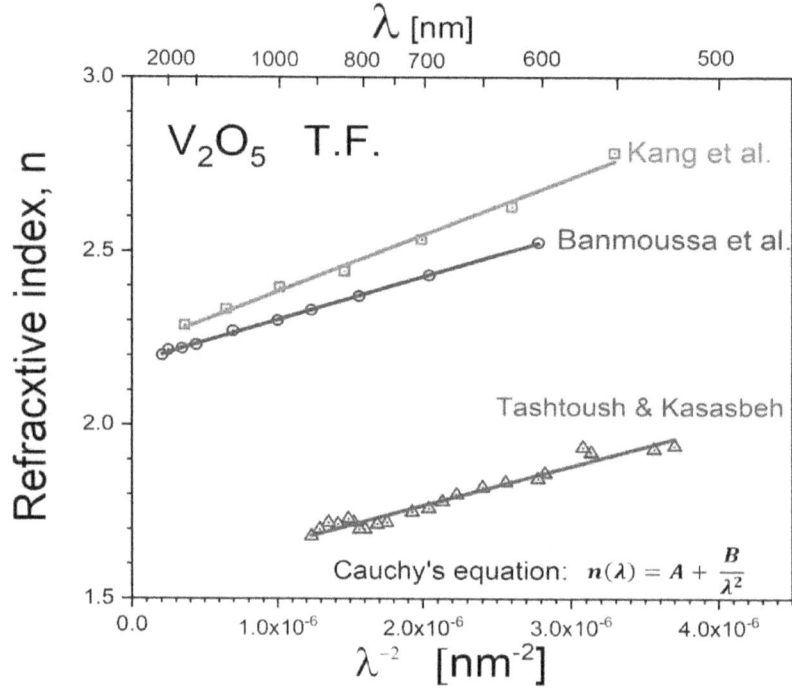

Fig. 3 Cauchy's plot of the refractive index (n) for V₂O₅ thin films [15, 38, 39].

The analysis of the dependence of the absorption coefficient on light frequency is very significant from the viewpoint of the semiconducting properties of vanadium oxide and its subsequent areas of application. Generally, the frequency dependence of the absorption coefficient ($\alpha(\omega)$) is rather different for various physical processes which occur during the interaction of light with the solid. In particular, the following cases can be observed [9]:

1.   Free carrier absorption
    a)   typical semiconductor $\alpha(\omega) \sim \omega^{-2}$ (7)

    b)   metals at low frequencies $\alpha(\omega) \sim \omega^{\frac{1}{2}}$ (8)

2.   Direct interband transition (conservation of crystal momentum)
    a)   allowed transition $\alpha(\omega) \sim \dfrac{(\hbar\omega - E_g)^{\frac{1}{2}}}{\hbar\omega}$ (9)

    b)   forbidden transition $\alpha(\omega) \sim \dfrac{(\hbar\omega - E_g)^{\frac{3}{2}}}{\hbar\omega}$ (10)

3.   Indirect interband transition (change in crystal momentum)
    a)   allowed transition $\alpha(\omega) \sim \dfrac{(\hbar\omega - E_g \pm \hbar\omega_{phonon})^2}{\hbar\omega}$ (11)

    b)   forbidden transition $\alpha(\omega) \sim \dfrac{(\hbar\omega - E_g \pm \hbar\omega_{phonon})^3}{\hbar\omega}$ (12)

The $\hbar\omega_{phonon}$ factor is generally omitted in Eqs (11) and (12) because of the fact that phonon energy is several times lower than the energy of electron transition. Eqs (9)-(12), known as Tauc equations, are applied to determine the band gap (E_g) of semiconductors.

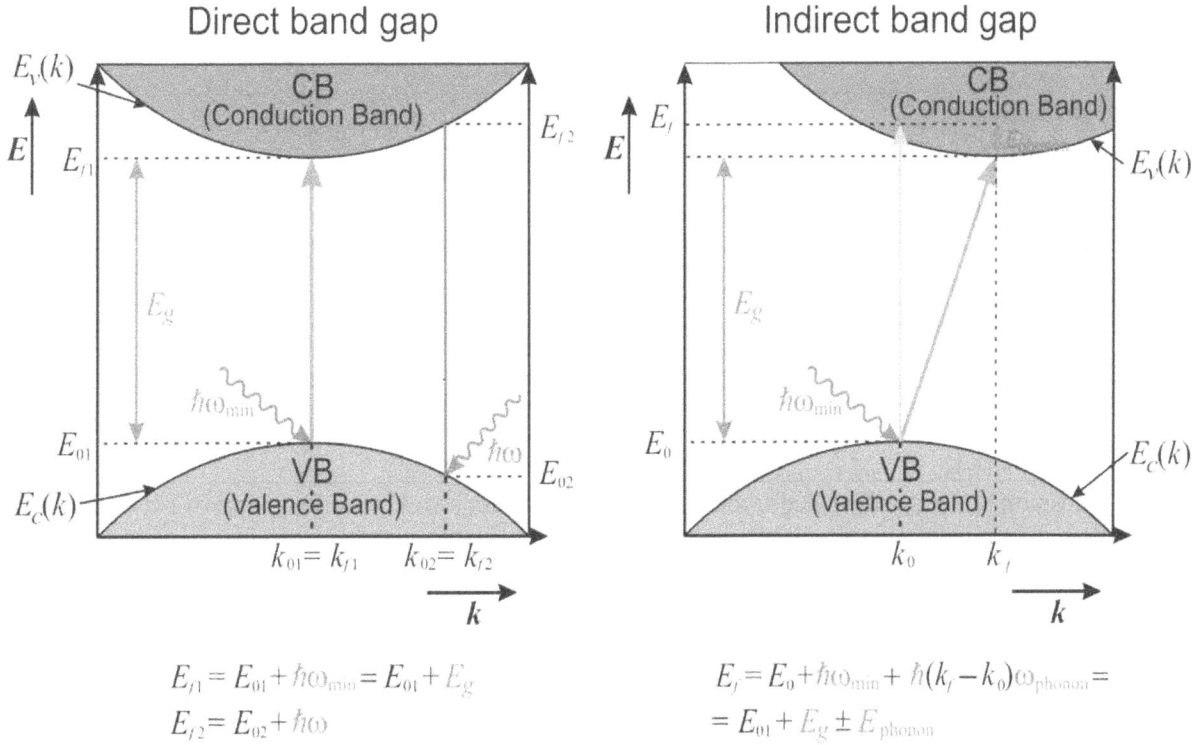

Direct band gap — Indirect band gap

$$E_{f1} = E_{01} + \hbar\omega_{min} = E_{01} + E_g$$
$$E_{f2} = E_{02} + \hbar\omega$$

$$E_f = E_0 + \hbar\omega_{min} + \hbar(k_f - k_0)\omega_{phonon} =$$
$$= E_{01} + E_g \pm E_{phonon}$$

Fig. 4 Direct and indirect mechanisms of electron interband transition

Fig. 4 shows the mechanisms of electron interband transition for direct and indirect semiconductors. A direct transition corresponds to the photon-electron interaction process in which the k-vector does not change. The crystal momentum of electrons and holes is the same in both the conduction band and the valence band. In an indirect transition photon, electron and phonon of the lattice take part. This process is accompanied by a change in the k-vector. The allowed transitions remain in agreement with particular selection rules, assuming a dipole model. On the other hand, if this model is not valid, the transition is called forbidden. More complex models can then be taken into account (involving, for instance, a magnetic dipole, electric quadrupole, etc.).

According to [35, 37], the edge is direct and forbidden. Diffuse reflectance spectra [41] give an E_g of 2.31 eV at room temperature, but the band edge has been determined to be direct and M. Mousavi et al. [42] observed that for V_2O_5 films prepared by means of spray pyrolysis E_g changes with the substrate temperature (T_{sub}). When T_{sub} increases, the E_g decreases gradually from 2.46 eV to 2.22 eV.

M. Kang et al. [4] studied the interband transition in a V_2O_5 film deposited via RF magnetron sputtering using absorption and photoluminescence spectral measurements. Transmission measurements indicate two distinct interband transitions, implying indirect and direct transitions.

2.2 Optical properties of V_2O_5 thin films – experimental results

Vanadium pentoxide thin films were deposited by means of reactive radio frequency sputtering. Deposition conditions of the thin films and their properties such as structure, morphology were described in detail elsewhere [43]. Table 2 summarizes characterization of the films used in the studies.

Table 2 Deposition conditions for a series of vanadium oxide thin films and their properties

Deposition conditions	Series			
	1	2	3	4
Ar flow [cm³/s]	6.3	6.7	6.7	6.7
O₂ flow [cm³/s]	0.3	0.7	2.1	2.0

Input power [W]	290	280	290	220
RF voltage (U_{rf}) [V]	1150	1150	1150	1350
Ar/O_2 gas atmosphere pressure [Pa]	$4.1 \cdot 10^{-2}$	$4.3 \cdot 10^{-2}$	$4.7 \cdot 10^{-2}$	$4.6 \cdot 10^{-2}$
Deposition time [min]	240	240	240 360	240
Thickness [nm]	420 ± 49	421 ± 43	431 ± 28 630 ± 42	396 ± 30
Substrate	Corning, Ti, fused SiO_2	Corning, Ti, fused SiO_2,	Corning, Ti, fused SiO_2, BVT	Fused SiO_2
Substrate temperature [K]	298	298	298	298

Optical transmittance and reflectance spectra were measured over a wide wavelength range from 180 to 3200 nm with a Lambda 19 Perkin-Elmer double beam spectrophotometer equipped with a 150 mm integrating sphere. Thin films from Series 1, 2 and 3 were the subject of spectrophotometric studies. Each of Figs 5-7 shows the reflectance (R), transmittance (T) and absorbance (A) spectra recorded for one of the thin films. The transparency region of vanadium pentoxide is limited by the fundamental absorption edge at ca. 500 nm. The reflectance values vary in the range of 10-20% for the Series 1 sample (mostly amorphous), in the range of ca. 0-10% for the Series 2 and 3 samples (crystalline). The observed non-monotonic plots of R, T and A may result from additional absorption bands due to the departure from stoichiometry [10, 11].

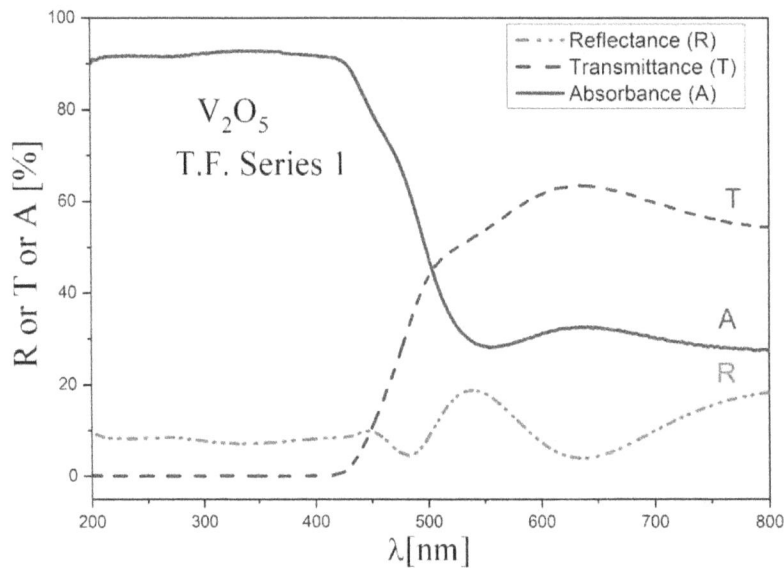

Fig. 5 Reflectance (R), transmittance (T) and absorbance (A) spectra recorded for a Series 1 V_2O_5 thin film

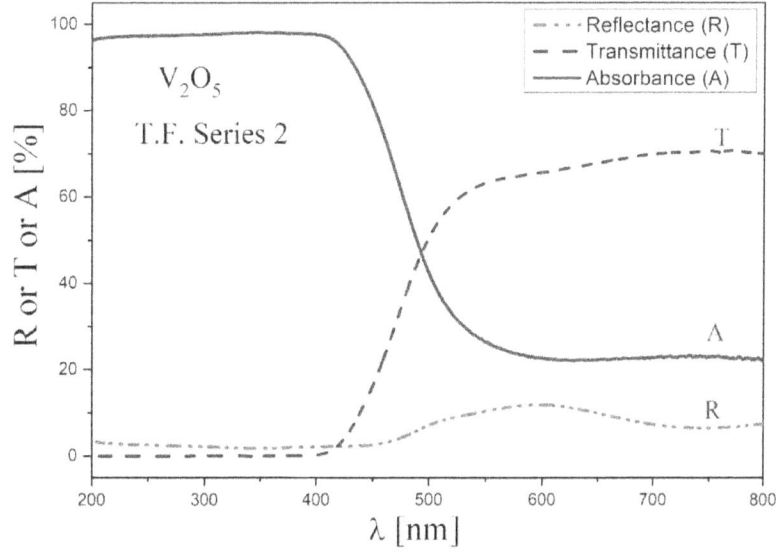

Fig. 6 Reflectance (R), transmittance (T) and absorbance (A) spectra recorded for a Series 2 V$_2$O$_5$ thin film

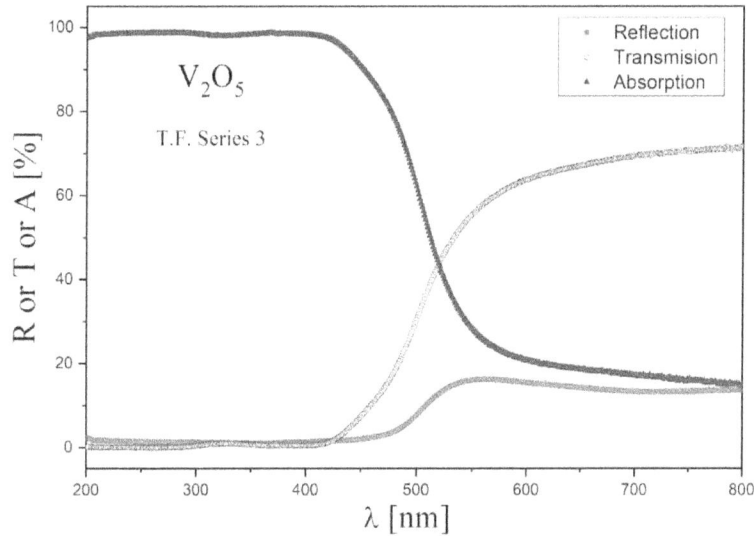

Fig. 7 Reflectance (R), transmittance (T) and absorbance (A) spectra recorded for a Series 3 V$_2$O$_5$ thin film

The absorption coefficient (α) and photon energy were determined from Figs 5-7 using the following equations:

$$\alpha = \frac{1}{d} \ln \frac{1-R}{T} \tag{13}$$

$$E_{photon}\,[eV] = \hbar\omega = \frac{1240}{\lambda\,[nm]} \tag{14}$$

where d represents film thickness.

One of the crucial parameters used to evaluate a semiconductor's properties is the band gap (E_g). The band gap of a semiconductor can be determined from experimentally measured transmittance T and reflectance R within the range of fundamental absorption using the following Tauc equation:

$$(\hbar\omega\alpha)^{\frac{1}{n}} = A(\hbar\omega - E_g) \tag{15}$$

where the A coefficient is constant and n, according to Eqs (9)-(12), assumes values ½, 3/2, 2 and 3 for direct allowed, direct forbidden, indirect allowed and indirect forbidden transitions, respectively.

Figs 8-11 illustrate the absorption coefficient data experimentally determined for the Series 2 thin film in the

71

coordinate system of $(\alpha\hbar\omega)^{1/n}$ versus $\hbar\omega$ for $n = \frac{1}{2}$, 3/2, 2 and 3, respectively.

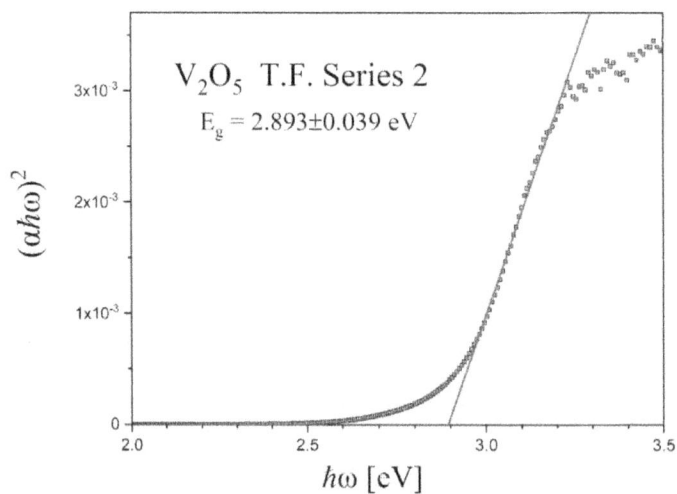

Fig. 8 Tauc plot ($n = 2$ – corresponding to a direct allowed transition) for a Series 2 V_2O_5 thin film

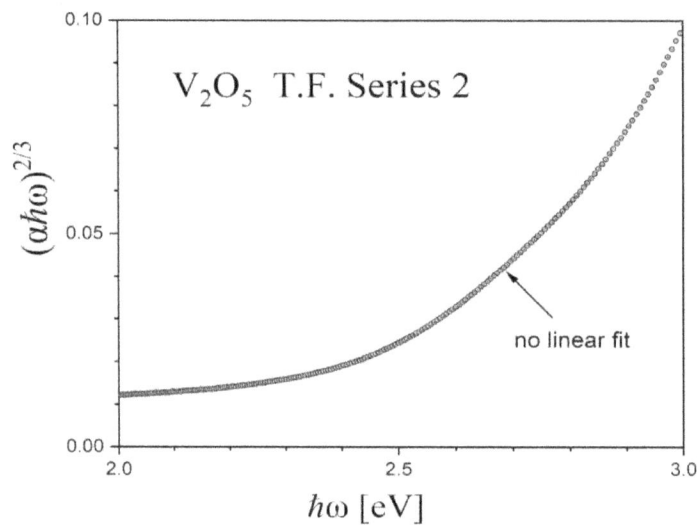

Fig. 9 Tauc plot ($n = 3/2$ – corresponding to a direct forbidden transition) for a Series 2 V_2O_5 thin film

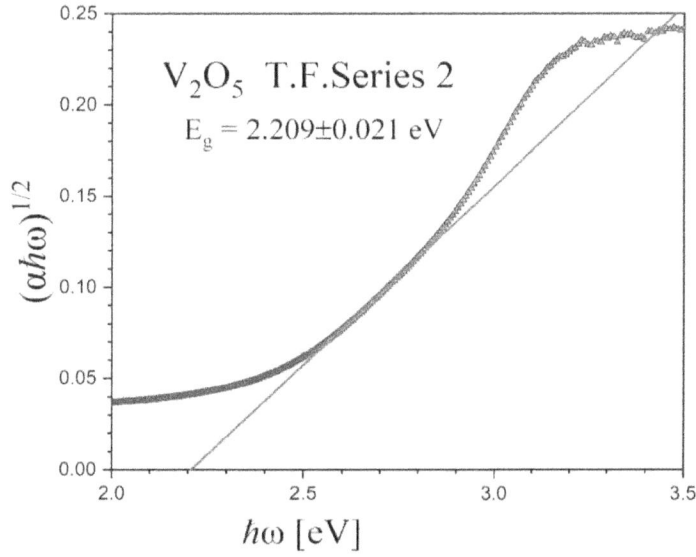

Fig. 10 Tauc plot ($n = 2$ – corresponding to an indirect allowed transition) for a Series 2 V_2O_5 thin film

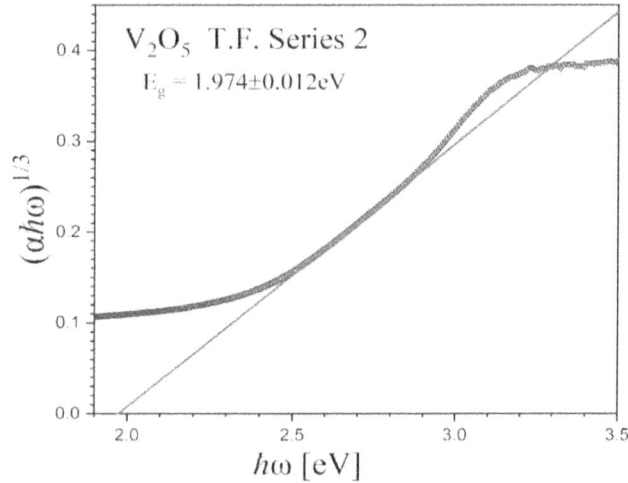

Fig. 11 Tauc plot ($n = 3$ – corresponding to an indirect forbidden transition) for a Series 2 V_2O_5 thin film

The band gap was determined by extrapolating the linear part of the best fit of $(\alpha\hbar\omega)^{1/n}$ vs. $\hbar\omega$ to $\alpha\hbar\omega = 0$. The values of E_g were obtained from the $\hbar\omega$ axis intercepts. Similar plots were computed for the other two V_2O_5 samples: Series 1 and Series 3. The results are presented in Table 3.

Table 3 Tauc plot results

Transition	Results		Series 1 T.F.	Series 2 T.F.	Series 3 T.F.
Direct Allowed	E_g [eV]		2.811 ± 0.065	2.893 ± 0.039	2.739 ± 0.089
	Linear regression	number of points (n)	29	36	31
		Pearson correlation	0.9983	0.9978	0.9958
Direct Forbidden	E_g [eV]		2.58 ± 0.11	2.307 ± 0.042	2.207 ± 0.051
	Linear regression	number of points (n)	20	36	99

		Pearson correlation	0.9968	0.9988	0.9948
Indirect Allowed		E_g [eV]	2.579 ± 0.064	2.209 ± 0.021	2.072 ± 0.030
	Linear regression	number of points (n)	32	37	120
		Pearson correlation	0.9981	0.9997	0.9976
Indirect Forbidden		E_g [eV]	1.969 ± 0.046	1.974 ± 0.012	1.782 ± 0.024
	Linear regression	number of points (n)	30	52	107
		Pearson correlation	0.9988	0.9998	0.9985

The analysis of spectrometric results presented in Figs 8-11 as well as in Table 2 suggests that V_2O_5 thin films undergo both direct and indirect transitions. However, it is not possible to decide which type of the electron interband transition is predominant in this case. Based on the following evidence:

- agreement with theoretical band calculations [23, 24, 25, 29]
- agreement with recent reports on single crystals [25, 44]
- agreement with recent experimental studies based not only on spectrophotometric measurements such as photoluminescence and ellipsometry [16]

the direct allowed (DA) transition can be considered the most probable. The available literature on the band gap of V_2O_5 thin film is vast. The impact of the following factors affecting the band gap of films based on vanadium pentoxide has been studied:

- thin film deposition technique [45, 46, 47] (Table 3),
- anisotropy [27, 28, 30, 48]
- film thickness [33]
- substrate type [48]
- non-stoichiometry [12, 49, 50]
- UV irradiation [49]
- chemical composition [51]
- temperature [16, 44, 45, 48]
- morphology [12, 29, 33, 46]

The results are presented in Table 4.

Table 4 Summary of findings concerning the band gap in vanadium pentoxide where DA, DF, IA and IF represent direct allowed, direct forbidden, indirect allowed and indirect forbidden transitions, respectively

Material	E_g [eV]	Electronic transition	Comments	Ref.
Single crystal	2.36 ‖a; 2.34 ‖c	DF	Anisotropy of E_g	[28]
Powder	2.31	DA		[48]
Single crystal	2.3	DA		[44]
Theoretical calculation	2.6	IA		[26]

Single crystal	theoretical 1.9 experimental 2.0	IA	ellipsometry	[52]
T.F. obtained via RF sputtering	2.15 $2.25^{p(O_2)=5\%}$ $2.37^{p(O_2)=20\%}$	IA DA	$d = 50$ nm $d = 1000$ nm effect of $p(O_2)$ during film deposition	[27]
T.F. obtained using electron beam	2.29-2.34	DF	Effect of UV irradiation (E_g increases after irradiation)	[49]
T.F. obtained using electron beam	2.32^{303K} 1.98^{603K}	DF	Effect of temperature deposition	[45]
Theoretical calculation	2.3^{DA}, 1.9^{IA} 2.3^{DA}, 2.1^{IA}	DA & IA	Bulk V_2O_5 Single-layer V_2O_5	[30]
Theoretical calculation	1.74 1.67 2.07		Bulk V_2O_5 Single-layer V_2O_5	[29]
T.F. obtained via magnetron sputtering	2.67^{20K}; 2.64^{300K} 2.26^{20K}; 2.16^{300K}	DA IA	$E_g = f(T)$	[16]
T.F. obtained via spray pyrolysis	1.98^{573K} 2.05^{673K}	DA	Effect of temperature deposition	[48]
T.F. obtained via: E-beam evaporation Magnetron sput. CVD Sol-gel Laser beam	 2.04-2.30 2.50^{RT}; 2.58^{673K} 2.16-2.59 2.15-2.20 2.42-2.49 2.20-2.50	- - - - - -	$E_g = f(p(O_2))$ $E_g = f(T_{deposition})$	[47]

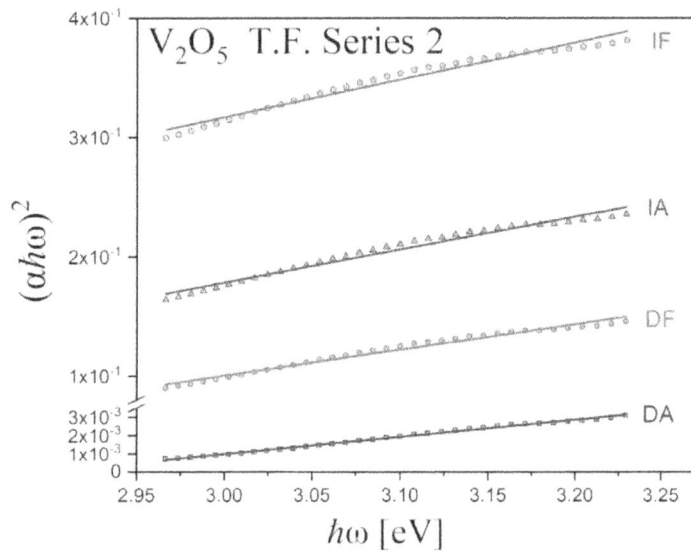

Fig. 12 Tauc plots for a Series 2 V_2O_5 thin film and the $\hbar\omega$ range of 2.97-3.23 eV

Fig. 12 illustrates the Tauc plots corresponding to $\hbar\omega$ energy between 2.97-3.23 eV. The best agreement with the theoretically predicted dependence is observed for the direct allowed (DA) transition. The results obtained by applying the least squares method are listed in Table 54.

Table 5 Results of calculations of the Tauc plot for the Series 2 V_2O_5 thin film and the ħω range of 2.97-3.23 eV, obtained using the least squares method.

	Intercept	ΔIntercept	Slope	ΔSlope	$R_{corel.}$	E_g [eV]	ΔE_g[eV]
DA	0.02684	0.00034	0.00928	0.00011	0.9977	2.89	0.04
DF	0.546	0.018	0.2156	0.0056	0.98905	2.53	0.15
IA	0.649	0.025	0.2758	0.0079	0.98702	2.35	0.15
IF	0.625	0.030	0.3138	0.0097	0.98478	1.99	0.16

2.3 Optical properties of V_2O_3

Based on papers [53-56], the most significant calculations were reported [57, 58]. The main controversy has been over the ordering of the components of the trigonally split t_{2g} band. Several band schemes have been suggested [59, 60, 61]. The V^{3+} ions in V_2O_3 have a $3d^2$ electronic configuration. These vanadium ions in the metallic phase (corundum) occupy two thirds of the octahedral sites formed by oxygen anions. A trigonal distortion causes the splitting of the t_{2g} orbital into a non-degenerate a_{1g} and a doubly degenerate e^{II}_g orbital [62]. Castellani et al. [57] suggested the formation of a molecular bond between the a_{1g} orbitals of V-V pairs spread into a band. This model is consistent with many experimental results [63-66]. However, taking into account the fact that the c/a lattice parameter ratio is too high for a_{1g} to couple between two V cations, Ivanov [67] and Ezhov et al. [68] contradicted the postulated model of the molecular orbital. Shinna et al. [69] assumed a strong hybridization of the V pair.

The electronic properties of V_2O_3 are strongly dependent on oxygen stoichiometry. A slight variation in oxygen concentration changes the effective mass [70, 71].

V_2O_3 is treated as the model system used to study the MIT in a correlated electron system (T_{MIT} = 160 K).

2.4 Optical properties of VO_2

The calculation of the electronic structure of VO_2 has been the subject of intensive research involving many models such as the cluster type [72, 73], tight-binding type [74-76], or augmented-plane-wave (APW) [77, 78], as well as energy band studies using Bloch functions in a linear combination of atomic orbitals [79].

Gavini et al. [80] determined the real (n) and imaginary (k) parts of the refractive index. Studies of the absorption coefficient performed by Gavini et al. and Merenda et al. [81] revealed that the electronic structure at E < 1.8 eV can be attributed to d-d transitions with a threshold at 0.6 eV. At 1.82 eV, the threshold for O2p-V3d transitions is observed, with peaks at 2.64 eV and 3.56 eV.

The temperature of MIT for bulk single-crystal VO_2 is 541 K [82]. Below T_{MIT}, VO_2 exhibits a monoclinic structure with the P2$_1$/c space group in which the partially filled d-band is split into an unoccupied part pushed past the π^* band and the filled part of the d-band. Above the T_{MIT}, VO_2 transforms to a tetragonal (rutile) phase with the partially filled d-band located at the Fermi level and the material is metallic [83]. Jiang et al. [84] studied the optical properties of vanadium dioxide thin films deposited under different oxygen partial pressures via reactive magnetron sputtering. The band gap decreased from 339.6 K to 319.4 K. The near-infrared extinction coefficient (k) and optical conductivity increased with decreasing oxygen partial pressure.

3. CONCLUSIONS

The electronic structure of the three main vanadium oxides (V_2O_3, VO_2 and V_2O_5) was reviewed. The optical properties of vanadium pentoxide thin films were determined. It was found that the direct allowed (DA) transition is the most probable type observed in the case of the studied films.

REFERENCES

1. N.B. Aetukuri, A.X. Gray, M. Drouard, M. Cossale, L. Gao, A.H. Reid, R. Kukreja, H. Ohldag, C.A. Jenkins,E. Arenholz, K.P. Roche, H.A. Durr, M. G. Samant, and S. P. Parkin, *Synthesis & characterization of nanostructure VO_2 thin film* ,Nat. Phys.9(10), (2013) 661-666

2. H.-P. Rust, J. Buisset, E.K. Schweizer, and L. Cramer, *Adsorption site determination with scanning tunnelling microscopy.* Rev. Sci. Instrum. 68 (1997) 129

3. N.F. Quackenbush, J.W. Tashman, J.A. Mundy, S. Sallis, H. Paik, R. Misra, J.A. Moyer, J.H. Guo, D.A. Fischer, J.C. Woicik, D. A. Muller, D. G. Schlom, and L. F. J. Piper, *Temperature dependence of the interband transition in a V_2O_5 film,* AIP Advances 3, 052129 (2013); Nano Lett.13(10), (2013) 4857-4861

4. M. Kang, S. W. Kim, Y. Hwang, Y. Um, J-W. Ryu, *Temperature dependence of the interband transition in a V_2O_5 film,* AIP Adv. 3 (2013) 052129

5. Hubbard, J. (1964). *Electron Correlations in Narrow Energy Bands. III. An Improved Solution*. Proceedings of the Royal Society of London. Series A, 281, 401-419

6. Peierls, R.E. (1955). *Quantum Theory of Solid*. London: Oxford University

7. Biermann, S., Poteryaev, A., Lichtenstein, A.I. & Georges, A. (2005). *Dynamical singlets and correlation-assisted peierls transition in VO2*. Physical Review Letters, 94, 026404 (4 pages)

8. Hohenberg, P., Kohn, W. (1964). *Inhomogeneous Electron Gas*, Physical Review, 136, B864-B871

9. M.S. Dresselhaus, in: Solid State Physics. Part II Optical properties of solids (2001)

10. R.D. Bringans, *The determination of the optical constants of thin films from measurements of normal incidence reflectance and transmittance*, J. Phys. D Applied Physics 10 (1977) 185501862

11. G. Lévêque, Y. Villachon-Renard, *Determination of optical constants of thin film from reflectance spectra*, Appl.Phys. 29 (1990) 3207-3212

12. M. Benmoussa, E. Ibnouelghazi, A. Bennouna, E.L. Ameziane, *Structural, electrical and optical properties of sputtered vanadium pentoxide thin films*, Thin Solid Films 265 (1995) 22-28

13. J.J. Ruiz-Perez, *Method for determining the optical constants of thin dielectric films with variable thickness using only their shrink reflection spectra*, J. Phys. D: Applied Physics 34 (2001) 2489-2496

14. A.K.S. Aquili, A. Maqsood, *Determination of thickness, refractive index, and thickness irregularity for semiconductor thin films from transmission spectra*, Appl. Optics 41 (2002) 218-224

15. N. M. Tashtoush, O. Kasasbeh, *Optical properties of vanadium pentoxide thin films prepared by thermal evaporation method*, Jordan.J. Phys. 6 (2013) 7-16

16. M. Kang, S. Won Kim, Y. Hwang, Y. Um, Ji-W. Ryu, *Temperature dependence of the interband transition in a V_2O_5 film*, AIP Adv. 3 (2013) 052129-1 - 05212.10

17. S.J. Petel, V. Kheraj, *Determination of refractive index and thickness of thin-film from reflectivity spectrum using genetic algorithm*, AIP Conference Proceedings 1536, 509 (2013); https://doi.org/10.1063/1.4810324

18. J. Szczyrbowski, A. Czapla, *On the determination of optical constants of films*, J. Phys. D: Appl. Phys. 12 (1979) 1737

19. J. Szczyrbowski, K. Schmalzbauer, H. Hoffmann, *Optical properties of rough thin films*, Thin Solid Films 130 (1985) 57-73

20. J.M. Bennett, E. Polletier, G. Albrand, J.P. Borgogno, B. Lazarides, C.K. Carniglia, R.A. Schmell, T.H.Allen, T. Tuttle-Hart, K.H. Guenther, A. Saxer, *Comparison of the properties of titanium dioxide films prepared by various techniques*, Appl. Opt. 28 (1989) 3303-3317

21. A. Brudnik, H. Czternastek, K. Zakrzewska, M. Jachimowski, *Plasma-emission-controlled d.c. magnetron sputtering of TiO_{2-x} thin films*, Thin Solid Films 199 (1991) 45-58

22. T. Pisarkiewicz, *Reflection spectrum for a thin film with non-uniform thickness*, J. Phys.D: Appl. Phys. 27 (1994) 160-164

23. W. Lambrecht, B. Djafari-Rouhani, M. Lannoo, J. Vennik, *The energy band structure of V_2O_5. I. Theoretical approach and band calculations*, J. Phys.C:Solid State Phys. 13 (1980) 2485-2501.

24. W. Lambrecht, B. Djafari-Rouhani, M. Lannoo, P. Clauws, L. Fiermans, J. Vennik, *The energy band structure of V_2O_5. II. Analysis of the theoretical results and comparison with experimental data*, J. Phys. C: Solid State Phys. 13 (1980) 2503-2517.

25. J.Y. Kempf, B. Silvi, A. Dietrich, C.R.A. Catlow, *Theoretical investigations of the electronic properties of vanadium oxides. 1. Pseudopotential periodic Hartree-Fock study of vanadium pentoxide crystal*, Chem. Mater. 5 (1993) 641

26. D.W. Bullett, *The energy band structure of V_2O_5: A simple theoretical approach*. J. Phys. C: Solid State Phys. 13 (1980) L595-L599

27. V. Eyert, K.-H. Höck, *Electronic structure of V_2O_5: Role of octahedral deformations*, Phys. Rev. B 57 (1998) 12727-12737

28. N. Kenny, C.R. Kennewurf, D.H. Whitmore, *Optical absorption coefficients of vanadium pentoxide single crystals*, J.Phys.Chem. Solids 27 (1966) 1237-1246

29. G. Stewart, *Electronic band structure of bulk and monolayer V_2O_5*, ghs9@case.edu (May 2, 2012) 1-26

30. A. Chakarbarti, K. Hermann, R. Druzinic, M. Witko, F. Wagner, M. Petersen, *Geometric and electronic structure of vanadium pentoxide: a density functional bulk and surface study*, Phys. Rev. B 16 (1999) 10583-10590.

31. R.A. Goschke, K. Vey, M. Maier, U. Walter, E. Göring, M. Klemm, S. Horn, *Tip induces changes of atomic scale images of the vanadium pentoxide surface*, Surf.Sci. 348 (1996) 305-310

32. A.D. Costa, C. Mathieu, Y. Barbaux, H. Poelman, G. Dalmai-Vennik, L. Lichtman, *Observation of the V_2O_5 (001) surface using ambient atomic force microscopy*, Surf. Sci. 370 (1997) 339-334

33. T.M. Tolhurst, B. Leedahl, J.L. Andrews, S. Banerjee, A. Moewes, *The electronic structure of ε'-V_2O_5: an expanded band gap in a double-layered polymorph with increased interlayer separation*, J. Mater. Chem. A 45 (2017) 23694-23703

34. D.S. Volzhenski, V.A. Grin, V.G, Savitskii, Kristallografiya 21 (1976) 1238

35. N. Kanay, O.R. Kennewurf, D.H. Whitmore, *Optical absorption coefficients of vanadium pentoxide single crystals* J.Phys. Chem. Solids 27 (1966) 1237-1246

36. V.G. Mokerov, Fiz. Tverd. Tela 15 (1973) 2393

37. Z. Bodo, I. Hevesi, *Optical absorption near absorption edge in V_2O_5 single crystal*, Phys. Stat. Solidi 20 (1967) K45

38. M. Benmoussa, E. Ibnouelghazi, A. Bennouna, E.L. Ameziane, *Structural, electrical and optical properties of sputtered vanadium pentoxide thin films*, Thin Solid Films 265 (1995) 22-28

39. M. Kang, I. Kim, S. Kim, H.Y. Park, *Metal-insulator transition without structural phase transition in V_2O_5 film*, Appl. Phys. Lett 98(2011) 131907-131916

40. F.A. Jenkins and H.E. White, *Fundamentals of Optics*, 4th ed., McGraw-Hill, Inc. (1981)

41. B. Karvaly, I. Hevesi, *Investigations of diffuse reflectance sopectra of V_2O_5 powder*, Z. Naturforsch. Teil A 26 (1971) 245-249.

42. M. Mousavi, A. Kompany, N. Shahtahmasebi, M.M. Bagheri-Mohagheghi, *Study of structural, electrical and optical properties of vanadium oxide condensed films deposited by spray pyrolysis technique,* Adv. Manuf. 1 (2013) 320-328

43. K. Schneider, *Vanadium oxides –properties and applications. Part IV,* this issue.

44. M.M. Margoni, S. Mathuri, K. Ramamurthi, R. Ramesh Babu, K. Sethuraman, *Sprayed vanadium pentoxide thin films: Influence of substrate temperature and role of HNO_3 on the structural, optical morphological and electrical properties*, Appl. Surf. Sci. 418 (2017) 280-290

45. C.V. Ramana, O.M. Hussain, B. Srinivasulu Naidu, P.J. Reddy, *Spectroscopic characterization of electron-beam evaporated* V_2O_5 *thin films,* Thin Solid Films 305 (1997) 219-226

46. E.E. Chain, O*ptical properties of vanadium dioxide and vanadium pentoxide thin films*, Appl. Optics 30 (1991) 2782.2787

47. S. Beke, *A review of the growth of* V_2O_5 *films from* 1885 *to* 2010, Thin Solid Films 519 (2011) 1763-1771

48. J. Meyer, K. Zilberberg, T. Riedl, A. Khan, *Electronic structure of vanadium pentoxide: An efficient hole injector for organic electronic materials*, J. Appl. Phys. 110 (2011) 033710 1-5

49. S. Krishnakumar, C.S. Menon, *Optical and electrical properties of vanadium pentoxide thin film*, Phys. Stat. Sol. (a) 153 (1996) 439-444

50. M.J. Scepanovic, M.Grujić-Brojčin, Z. Dohčević-Mitrović, K. Vojisavljevic, Z. Popovic, *The effects of nonstoichiomnetry on optical properties of oxide nanopowders*, Acta Phys. Polon, Ser a 112(2007) 1013

51. D. Souri, K. Shomalian, *Band gap determination by absorption spectrum fitting method (ASF) and structural properties of different compositions of* $(60-x)V_2O_{5.4}0TeO_{2.x}Sb_2O_3$ *glasses*, J. Non-Cryst.Solids 355 (2009) 1597-1601

52. J.C. Parker, D.J. Lam, Y.-N. Xu, W.Y. Ching, *Optical properties of vanadium pentoxide determined from elipsometry and band- structure calculations*, Phys.Rev. B 42(1990) 5289-5293

53. I. Nebenzahl, M. Weger, *Band structure of the 3d-t_{2g} sub-shell of* V_2O_3 *and* Ti_2O_3 *in the tight-binding approximation,* Philos.Mag. 24(1971) 1119

54. M. Weger, *Application of the excitonic model to the metal-insulator transition of a simple band model,* Philos. Mag. 24(1971) 1095

55. J. Ashkenazi, M. Weger, *A model for the metal-to-insulator transition in* V_2O_3, Adv. Phys. 22(1973) 207-261

56. J. Ashkenazi, T. Chuchem, *Band structure of* V_2O_3 *and* Ti_2O_3, Philos.Mag.32 (1975) 763

57. J. Ashkenazi, M. Weger, *The effect of band structure and electron-electron interactions on the metal-insulator transitions in* Ti_2O_3 *and* V_2O_3, J. Phys. 37 (1976) C4-189 –C4-180

58. C. Castellani, C.R. Natoli, J. Ranninger, Phys. Rev. *Magnetic structure of* V_2O_3 *in the insulating phase*, B 18 (1978) 4945-4967

59. S. Yu Ezhov, V.I. Ansimov, D.I. Khomski, G.A. Sawatzky, *Orbital occupation, local spin and exchange interactions in* V_2O_3, Phys. Rev. Lett 83 (1999) 4136

60. J.M. Honig, L.L. Van Zandt, R.D. Board, H.E. Weaver, *Study of* V_2O_3 *by X-ray photoelectron spectroscopy*, Phys.Rev. B6 (1972)323.

61. P. Shuker, Y. Yacoby, *Differential reflectance spectra and band structure of* V_2O_3 Phys. Rev. B 14 (1976) 2211-

62. J.W. Taylor, T.J. Smith, K.H. Andersen, H. Capellman, R.K. Kremer, A. Simon, K-U Neumann, K.R.A. Ziebeck, *Spin-spin correlations in the insulating and metallic phases of the Mott system* V_2O_3, Eur. Phys. J. B 12 (1999) 199-207

63. L. Paolasini, C. Vettier, F. De Bergevin, F. Yakhou, D. Mannix, W. Neubeck, A. Stunault, M. Altarelli, M. Fabrizio, P.A. Metcalf, and J.M. Honig. *Direct observation of orbital in* V_2O_3 *by X-ray resonant technique,* Physica B, 281-282 (2000) 485-486.

64. M. Fabrizio, M. Altarelli, and M. Benfatto, *X-ray resonant scattering as a direct probe of orbital ordering in transition- metal oxides,* Phys. Rev. Lett., 80 (1998) 3400.

65. L. Paolasini, C. Vettier, F. De Bergevin, F. Yakhou, D. Mannix, A. Stunault, W. Neubeck, M. Altarelli, M. Fabrizio, P.A. Metcalf, and J.M. Honig., *Orbital occupancy order in: resonant X-ray scattering results,* Phys. Rev. Lett., 82(1999) 4719.

66. W. Bao, C. Broholm, G. Aeppli, P. Dai, J.M. Honig, and P. Metcalf. *Dramating switching of magnetic exchange in a classic transition metal oxides: evidence for orbital ordering,* Phys. Rev. Lett., 78 (1997) 507.

67. V.A. Ivanov. *The tight-binding approach to the corundum-structure d compounds,* J. Phys.: Condens. Matter, 6 (1994) 2065-2076.

68. S.Yu. Ezhov, V.I. Anisimov, D.I. Khomskii, and G.A. Sawatzky. *Orbital occupation, local spin, and exchange interactions in V_2O_3,* Phys. Rev. Lett., 83 (1999) 413 6-4139.

69. R. Shiina, F. Mila, F.-C Zhang, and T. M. Rice, *Atomic spin, molecular orbitals, and anomalous antiferromagnetism in insulating V_2O_3* Phys. Rev. B, 63 (2001) 144422.

70. C.A. Carter, T.F. Rosenbaum, P, Metcalf, *Mass enhancement and magnetic order at the Mott-Hubard transition,* Phys.Rev. B 48 (1993) 16841-16844.

71. J. Trastoy, Y. Kaicheim, J. del Valle, I. Valmanski, I.K. Schuller, *Electronic materials,* J. Mater.Sci. 53 (2018) 9131-9137.

72. N.I. Lazukova, V.A. Gubanov, *Metal-insulator phase transition in VO_2,* Solid State Commun.20 (1976) C4-59.

73. C. Sommers, R. de Groot, D. Kaplan, A. Zylberstejn, Cluster calculations of the electronic d-states in VO_2, J. Phys (Paris) Lett. 36 (1975) L-157 -160.

74. T.K. Mitra, S. Chatterjee, G.J. Hyland, *A LCOAO approach to the band structure of rutile VO_2,* Phys.Lett. A37 (1971) 221-222.

75. T.K. Mitra, S. Chatterjee, G.J. Hyland, *A method for obtaining parametrized bands of rutile VO_2,* Can.J.Phys. 51 (1973) 352-365

76. T. Altanham, G.J. Hyland, *One-electron dispersion relations in metallic VO_2,* Phys.Lett. A 61 (1977) 426-428.

77. E. Caruthers, L. Kleinman, H.I.Zhang, *Energy bands of metallic VO_2,* Phys. Rev. B 7 (1973) 3753-3760

78. E. Caruthers, L. Kleinman, *Energy bands of semiconducting VO_2,* Phys. Rev. B 7 (1973) 3760-3766

79. M. Gupta, A.J. Freeman, D.E. Elilis, *Electronic structure and lattice stability of metallic VO_2,* Phys. Rev. B. 16 (1977) 3338-3351

80. A. Gavini, C.C.Y. Kwan, *Optical properties of semiconducting VO_2 films,* Phys.Rev. B 3 (1972) 3138-3143

81. P. Merenda, D. Kaplan, C. Sommers, *Near band gap optical absorption on semiconducting VO_2,* J.Phys. Coll0ques 37 (1976) C4-59 –C4-62

82. F.I. Morin, *Oxides which show a metal-to-insulator transition at the Neel temperature,* Phys.Rev. Lett. 3 (1959) 34-36

83. T.C. Koethe, Z. Hu, M.W. Haverkort, C. Schüßler-Langeheine, F. Venturini, N.B. Brookes, O.Tjernberg, W. Reichelt, H.H. Hsieh, H-J. Lin, C.T. Chen, L.H. Tjeng, *Transfer of spectral weight and symmetry across the metal-insulator transition in VO_2,* Phys.Rev.Lett. 97 (2006) 116402 (4pp)

84. M. Jiang, Y. Li, S. Li, H. Zhou, X. Cao, S. Bao, Y. Gao, H. Luo, P. Jin, *Room temperature optical constants and band gap evolution of phase pure M_1-VO_2 thin films deposited at different oxygen partial pressure by reactive magnetron sputtering,* J. Nanomater. 2014 (2014) 183954 (6 pp)

VI. Defect structure and electrical properties of vanadium oxides

Abstract

The point defect structure of V_2O_3, VO_2 and V_2O_5 is reviewed. VO_2 and V_2O_5 thin films were deposited by means of RF sputtering from a metallic V target in a reactive $Ar+O_2$ atmosphere. X-ray diffraction (XRD), Rutherford backscattering (RBS) and secondary-ion mass spectrometry (SIMS) were used to determine the chemical and phase composition as well as the profile distribution of elements in as-sputtered and hydrogen-treated VO_2 films. The electrical properties of the V_2O_5 thin films were determined by means of impedance spectroscopy. At elevated temperatures thin films was fpund to interact with oxygen. Both singly and doubly ionized oxygen vacancies and electrons were the product of these interactions. The chemical diffusion coefficient was determined by measurements of transient electrical conductivity.

1. INTRODUCTION

Vanadium oxides exhibit fascinating electrical, magnetic, optical and catalytic properties. This stems from the large number of possible vanadium oxidations states – ranging from V^{2+} to V^{5+} – as well as the occurrence of the metal-insulator transition (MIT). As Morin noted, resistance in some vanadium oxides, such as dioxide (VO_2), sesquioxide (V_2O_3) and pentoxide (V_2O_5), increases by several orders of magnitude when temperature decreases from high to low across the transition temperature (T_{MIT}) [1]. This MIT makes vanadium oxides attractive candidates for various technological applications such as field-effect devices, terahertz materials, memory devices, metamaterials, ultrafast switching devices, infrared detectors, thermal sensors, smart windows, electrodes in lithium ion batteries, energy storage materials [2] and chemical sensors [3].

However, vanadium oxide properties are strongly affected by various point defects originating from deviations from stoichiometry and the presence of foreign ions unintentionally or intentionally added during doping or as impurities. One of the significant challenges with regard to this group of materials is therefore elucidating the defect chemistry involving the types and concentrations of point defects and their impact on the properties of a given vanadium oxide.

2. DEFECT STRUCTURE OF V_2O_3

Vanadium sesquioxide (V_2O_3) undergoes the transition from an insulator to a metal (MIT) at 165 K. Little is thus known about the defect structure of undoped V_2O_3 below T_{MIT}.

B. Sass et al. [4] studied thin films of V_2O_3 which had a thickness of 4–300 nm and were grown on oriented sapphire substrates via reactive DC magnetron sputtering. Below 160 K, resistivity increases steeply by nearly five orders of magnitude, indicating a phase transition into the insulating state. The transition is fully reversible and exhibits hysteresis, revealing that it is a first-order one. The band gap determined at 77 K was 0.45 eV. While in the insulating phase, V_2O_3 films are antiferromagnetically ordered.

V. Simic-Milosevic et al. [5] investigated the electronic structure of non-stoichiometric V_2O_3 films grown on Au

(111) by means of scanning tunneling microscopy and spectroscopy. They found that the band gap varies between 0.2 and 0.75 eV depending on the sample preparation conditions. The changes in gap size are attributed to local deviations from the ideal V_2O_3 stoichiometry in films produced at more oxidizing or reducing conditions. The films exhibit p-type conductivity. Doping experiments with various cations, such as Ti, Cr, Fe, Zr, Al, and Mg, demonstrated the possibility of modifying the electronic properties of V_2O_3. It was shown that the substitution of V with Ti stabilizes the metallic phase, while the insulating phase is supported by the other dopants.

D. Wickramaratane et al. [6] investigated the impact of native point defects in the form of Frenkel defects on the structural, magnetic and electronic properties of VO_2 and V_2O_3, using first-principles density functional theory (DFT) calculations. They found small polarons to be stable, either accompanied by defects or as a self-trapped species. Self-trapped small polarons in V_2O_3 can lead to hopping-like conductivity with a migration barrier of 0.09 eV for small hole polarons and 0.11 eV for electron polarons.

The temperature range of the insulating phase of V_2O_3 can be shifted towards higher temperatures via doping. The most effective dopant is chromium. Doping with Cr has a marked effect on the lattice parameters and MIT – it shifts the latter to room temperature [7].

Above T_{MIT}, vanadium sesquioxide exhibits a metal deficit ($V_{2-y}O_3$). The oxygen pressure dependence within 1400-1700 K of y nonstoichiometry can be expressed by $y \propto p_{O2}^{3/4}$ [8]. The dependence of nonstoichiometry (y) on oxygen partial pressure over the temperature range of 1400-1700 K can be expressed as $y \propto p_{O2}^{3/4}$ [8]. This dependence may be represented as follows:

$$\frac{3}{2}O_2 \leftrightarrow 2V_{Va}^x + 3O_O \qquad (1)$$

where V_{Va}^x denotes neutral vanadium vacancies (according to the Kröger-Vink notation).

The Kröger-Vink point defect notation is employed in this equation with one modification. According to this notation, vacancies are denoted by V, but this coincides with the chemical symbol of vanadium, which is why the symbol V_V might be interpreted either as a vanadium atom in the vanadium site, or as a vanadium vacancy. In order to avoid this ambivalence, vanadium vacancies are represented using V_{Va} both in Eq. (1) and elsewhere in the present work. The application of the law of mass action yields:

$$y = [V_{Va}^x] = K_V^{1/2} \, p_{O2}^{\frac{3}{4}} \qquad (2)$$

where K_V is the equilibrium constant of the reaction (8). Based on the thermogravimetric results obtained by Wakihara and Katsura [8], the equilibrium constant (K_V) may be determined from Eq. (2):

$$K_V = 3.185 \cdot 10^{-11} \exp\frac{546.8 \text{ kJ/mol}}{RT} \qquad (2a)$$

Taking into account Eqs (2) and (2a), the dependences of the nonstoichiometry (y) in $V_{2(1-y)}O_3$ on temperature and oxygen partial pressure were illustrated in Fig. 1 and Fig. 2, respectively. Under a constant oxygen pressure, this nonstoichiometry decreases with increasing temperature. This means that the enthalpy of the formation of vanadium vacancies is negative. Assuming that the defect structure is described by Eq. (1) and Eq. (2) the electroneutrality condition can be expressed as follows:

$$n = p \qquad (3)$$

where n and p denote the concentration of electrons and electron holes, respectively.

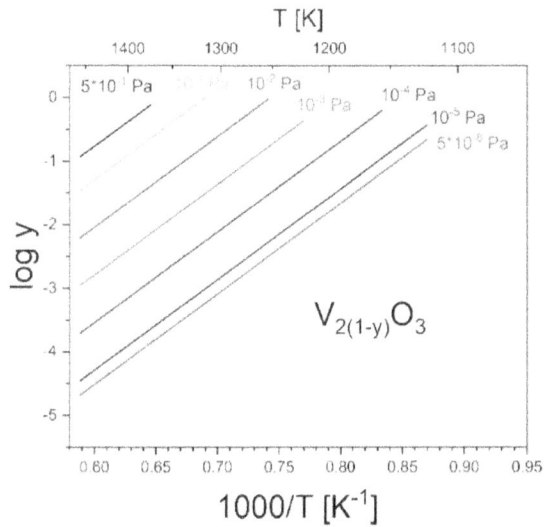

Fig. 1 Deviation from stoichiometry (y) in
$V_{(2-y)}O_3$ vs. temperature.

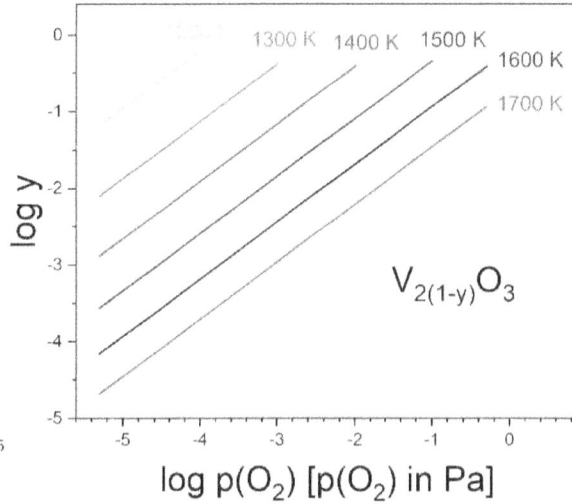

Fig. 2 Deviation from stoichiometry (y) in
$V_{(2-y)}O_3$ vs. $p(O_2)$.

3. DEFECT STRUCTURE OF VO₂ THIN FILMS

3.1 Literature survey

Vanadium dioxide (VO_2) has attracted great interest because of the fact that its metal-insulator transition occurs near room temperature ($T_{MIT} = 341$ K). However, the preparation of stoichiometric VO_2 remains a great challenge, because the vanadium-oxygen system is highly complex due to the relatively high accessibility of different vanadium oxidation states and their high tolerance for point defects. The temperature of the metal-insulator transition may be affected by deviations from the stoichiometric composition ([O]/[V] ≠ 2, where [O] and [V] denote the number of atoms in the chemical formula, respectively) and as a result of doping with foreign cations. The afore-mentioned deviations may stem from preparation or heating conditions. H. Kim et al. [9] studied VO_2 epitaxial thin films deposited on single-crystal sapphire substrates by means of pulsed laser deposition. They observed that the T_{MIT} increased with oxygen partial pressure, $p(O_2)$, during deposition. This is illustrated in Fig. 3.

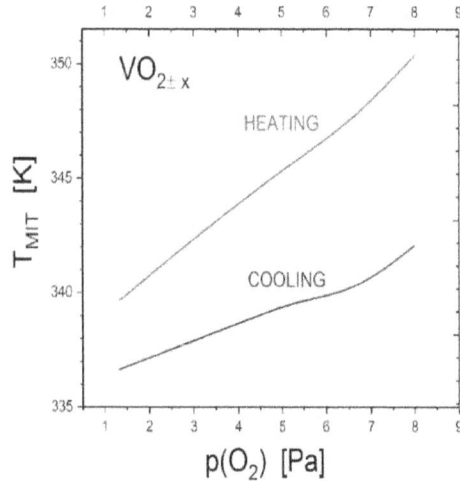

Fig. 3 Dependence of T_{MIT} on oxygen partial pressure during thin film deposition [9].

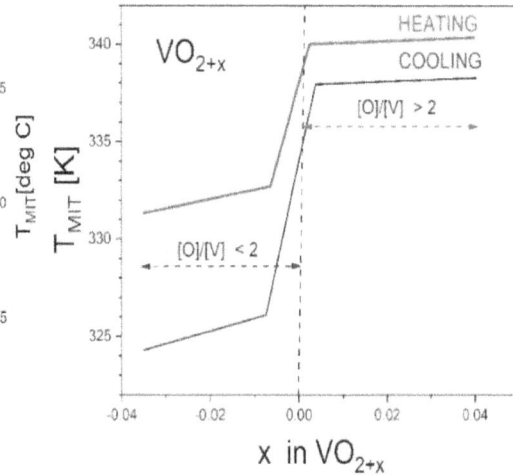

Fig. 4 T_{MIT} vs. stoichiometry [10].

Fig 4 illustrates the dependence of T_{MIT} on the [O]/[V] ratio, as reported by C. Blaauw et al. [10]. As the [O]/[V] ratio increases, so does the transition temperature. Yu et al. [11] prepared VO_2 films on silicon using direct-current magnetron sputtering in an argon-oxygen atmosphere. They found that the hysteresis loop across the MIT was narrowed for a high oxygen/argon ratio because the defects introduced by excess oxygen reduced the energy of the phase transition.

However, the exact kind of point defects resulting from deviations from stoichiometry is still an open issue. There is a consensus that for [O]/[V] < 2, oxygen vacancies (V_O) are the predominant point defects [12]. On the other hand, for [O]/[V] > 2 some authors [13] claim the presence of oxygen interstitials (O_i), whereas others insist on vanadium vacancies (V_{Va}). The remaining researchers are of the opinion that both oxygen interstitials and vanadium vacancies are present [14-16]. Moreover, the presence of hydrogen defects (also called 'protonic defects') was also postulated [17, 18]. These defects may form in VO_2 samples obtained via the hydrogen reduction of V_2O_5 or the interaction of VO_2 with water vapor [11].

3.2 Experimental results

Schneider et al. [19] prepared VO_2 thin films by means of RF sputtering from a metallic V target in a reactive Ar+O_2 gas flow controlled atmosphere. The as-sputtered thin films were annealed for 3 h in a hydrogen atmosphere at 573 K. X-ray diffraction (XRD), Rutherford backscattering spectrometry (RBS) and secondary-ion mass spectrometry (SIMS) were used to determine the chemical and phase composition as well as the profile distribution of elements in the as-sputtered and hydrogen-treated films. Fused silica, silicon, Corning and alumina served as a substrate as well as the controlled substrate temperature (T_{sub} = 298 K). Prior to the sample deposition, the target was pre-sputtered in an Ar+O_2 atmosphere in order to stabilize the sputtering conditions (voltage, pressure, gas composition) and to equilibrate the surface of the oxidized target. XRD analysis revealed that the as-sputtered thin films were amorphous. The deposited thin films were reduced for 3 h in a hydrogen gas atmosphere (p_{H2} = 10^5 Pa) at 573 K.

RBS measurements were performed at the Institute of Nuclear Physics of the Goethe University Frankfurt, using a 7 MeV $4He^+$ ion beam with a 171° backscattering angle. The details concerning the applied experimental conditions had been described elsewhere [20]. For data evaluation, the SIMNRA software [21] was used; the electronic stopping power data by Ziegler and Biersack, Chu and Yang's theory for electronic energy-loss straggling calculations, and Andersen's function dedicated to screening to Rutherford cross-section were applied. The contribution from double and/or multiple scattering into the RBS spectra [22] was taken into account using the computational features of SIMNRA.

The O/V ratio determined from RBS experiments was 2.030. The nominal composition was thus either $V_{1.000}O_{2.030}$

or $V_{0.985}O_2$ depending on whether the oxygen excess or the vanadium deficit vs. the stiochiometric composition is emphasized.

The observed deviation from stoichiometry could have originated from either of the following defect equilibria:

- $O_O \Leftrightarrow \frac{1}{2}O_2 + V_O^{\bullet\bullet} + 2e'$ (4)

- $nil \Leftrightarrow e' + h^{\bullet}$ (5)
- $nil \Leftrightarrow V_{Va}^{''''} + 2V_O^{\bullet\bullet}$ (6)
- $O_O \Leftrightarrow O_i'' + V_O^{\bullet\bullet}$ (7)

None of these reactions alone is sufficient to explain the obtained RBS results. There are two possibilities. The first one is based on the assumption that protonic defects are formed. Iwahara et al. [23] discovered high-temperature proton conductivity in oxides that form oxygen vacancies. The protonic defects are the result of interactions involving either gaseous hydrogen or water vapour [18]:

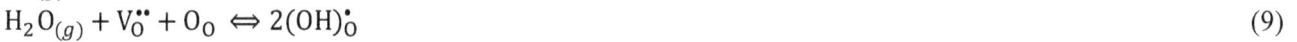

$$H_{2(g)} + 2O_O \Leftrightarrow 2(OH)_O^{\bullet} + 2e'$$ (8)
$$H_2O_{(g)} + V_O^{\bullet\bullet} + O_O \Leftrightarrow 2(OH)_O^{\bullet}$$ (9)

Another way to explain the RBS results is by assuming that either equilibria (4)-(6) or (4), (5) and (7) play a role. In order to decide which case is more likely, the SIMS experimental method was used.

Secondary-ion mass spectroscopic (SIMS) investigations were performed at the Institute of Materials Science of Technische Universität Darmstadt, using $^{133}Cs^+$ primary ions; positive secondary ions were collected and analyzed with a CAMECA ims 5f equipment with a base pressure of $3 \cdot 10^{-8}$ Pa.

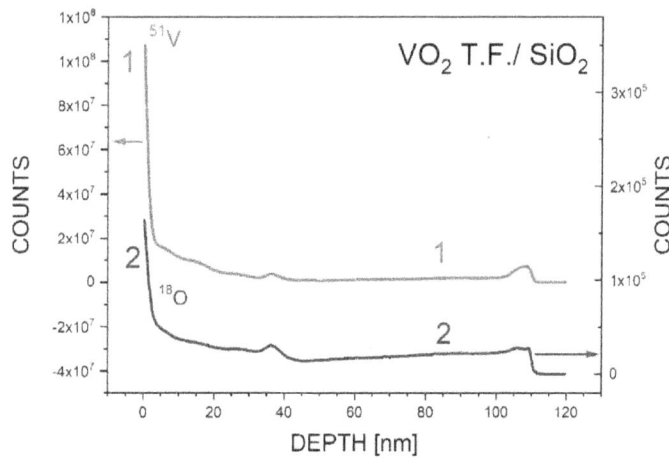

Fig. 5 SIMS concentration profile of vanadium and oxygen

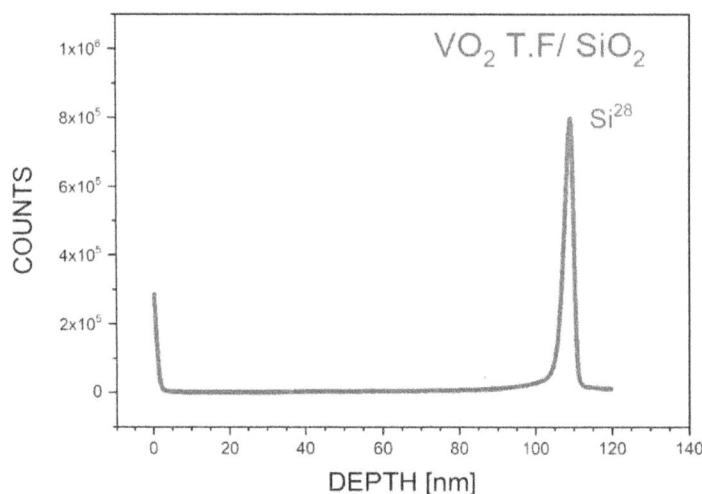

Fig. 6 SIMS concentration profile of silicon

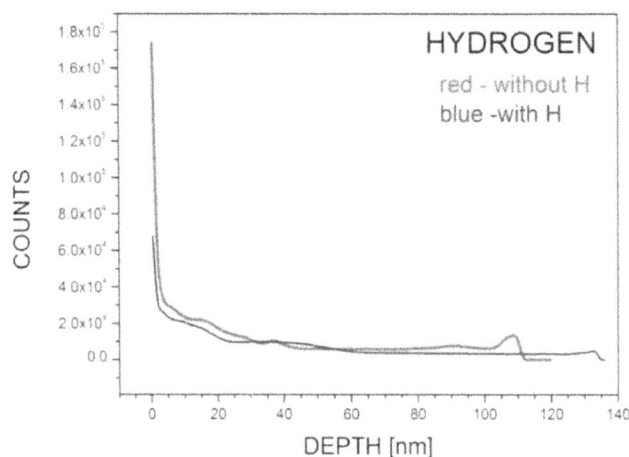

Fig. 7 SIMS concentration profile of hydrogen

Figs 5-7 present the concentration profiles of vanadium and oxygen, silicon, and hydrogen, respectively. These figures illustrate the profiles for as-sputtered thin films and such profiles of the hydrogen treated. The high peaks observed at depths in the range of 0-10 nm might have resulted from the accumulation of elements at the surface and near-surface layers due to interface reactions such as chemisorption and segregation.

No diffusion between the film material and the applied substrate was observed (Fig. 6). Hydrogen treatment did not cause the accumulation of this element in the film material (Fig. 7). The presented SIMS results indicate that the hypothesis about the occurrence of protonic defects can be discounted. Likewise, the formation of oxygen ion interstitials (Eq. 7) suggested by some authors [24, 25] can be rejected taking into account the ion radii values – r_{V4+} =58 pm and r_O =136 pm [26]. Oxygen interstitial defects were found in metal oxides such as uranium dioxide (UO_{2+x}) [27], which contain larger metal ions.

4. ELECTRICAL PROPERTIES & DEFECT STRUCTURE OF V_2O_5 THIN FILMS

4.1 Experimental

The electrical properties of the investigated films were determined by means of impedance spectroscopy. The applied measurement setup allowed the temperature and gas atmosphere composition to be controlled. A frequency range between 0.1 Hz and 1.4 MHz was covered; the amplitude was 10 mV, and the temperature ranged from 420 K to 670 K. The measurements were performed with a Solartron system (1260 Frequency Response Analyzer (FRA) + 1294 dielectric interface). The experimental parameters and data acquisition were controlled by the software bundled with the FRA. The impedance spectra were analyzed using ZView software.

The measurements were performed with a conductometric sensor substrate provided by BVT Technologies – CC1.W (Fig. 8). The atmosphere of the sample chamber was a mixture of artificial air and argon. The flow rates of gases were independently controlled using an MKS Type 1179A mass-flow controller. The total flow rate was maintained at the same level of 100 sccm.

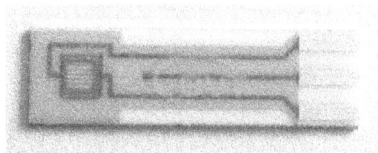

Fig. 8 Substrate used for EIS experiments.

4.2 Electrical conductivity vs. T

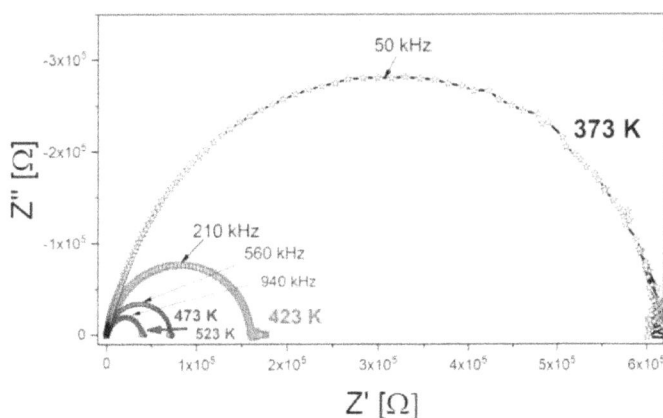

Fig. 9 Nyquist plots recorded for the V_2O_5 thin film at temperatures ranging from 373-523 K.

Fig. 9 shows the impedance spectra recorded at lower temperatures (373-523 K), presented on a complex Z'' vs. Z' plane (Nyquist plot). The plots can be interpreted in terms of an equivalent circuit composed of a resistor and a non-Debye constant phase element (CPE) connected in parallel. An increase in temperature is accompanied by a decrease in ohmic resistance, which is a behaviour typical of semiconducting materials.

Entirely different impedance spectra were observed at higher temperatures (above 528 K). An example is presented in Fig. 10.

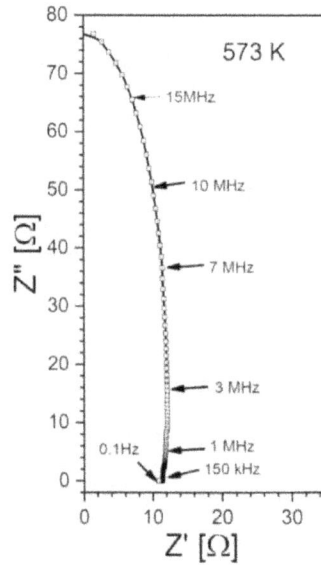

Fig. 10 Nyquist plots recorded for the V_2O_5 thin film at 573K.

In this case, the complex component of impedance (Z'') assumes positive values, which indicates a non-negligible contribution of inductance. The presented spectrum can be interpreted using an equivalent circuit composed of a resistor (R) and an inductor (L) connected in series. Fig. 11 illustrates the Nyquist plots on a complex admittance plane at 573 K, 623 K and 673 K. The lack of experimental points above the Y' axis suggests that the contribution of capacitance elements was negligible.

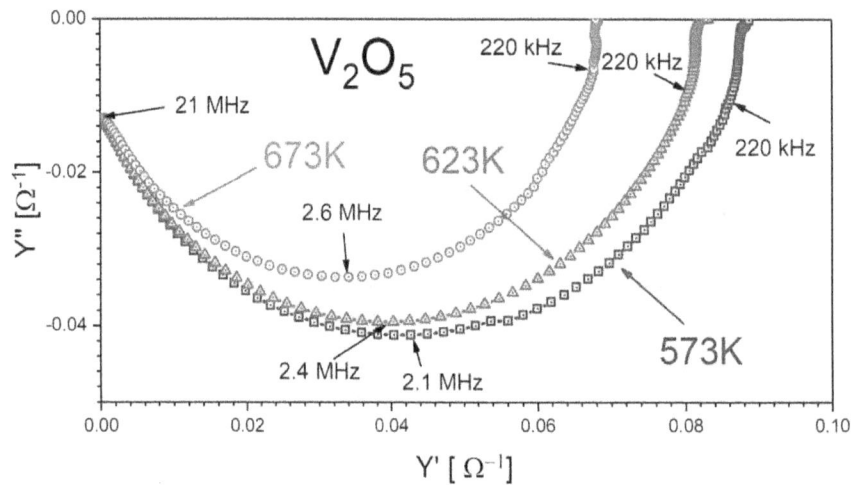

Fig. 11 Experimental spectra recorded at 573, 623 and 673 K, presented on a complex admittance plane.

The values of inductance (L) can be determined from the dependence of the imaginary component of impedance (Z'') on angular frequency (ω):

$$Z'' = j\omega L \tag{10}$$

where j is the imaginary unit. Fig. 12 illustrates this dependence at 573 K.

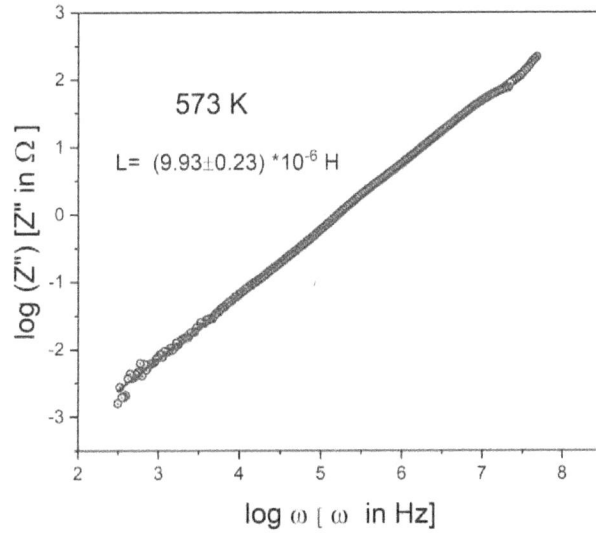

Fig. 12 Dependence of Z'' on ω.

Fig. 13 shows the experimental admittance spectrum (points) and theoretical dependence corresponding to the postulated equivalent circuits – L-R (model 1) and L-R-CPE (model 2). Good agreement is observed for the lower frequencies (0-0.32 MHz) and to frequencies from the range of 4.20-4.45 MHz in the case of model 1, whereas assuming the L-R-CPE equivalent circuit (model 2) the agreement is very good.

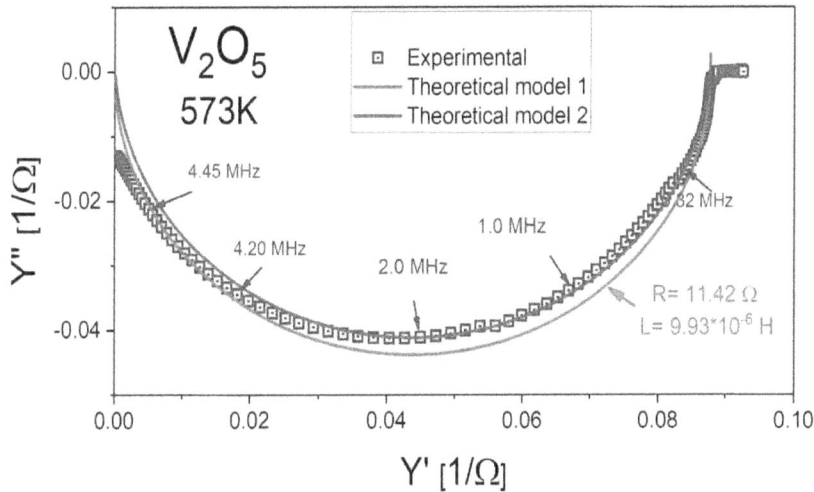

Fig. 13 Admittance spectrum recorded for the thin film at 573 K; points represent experimental data and lines are theoretical dependences: the red one (1) corresponds to the L-R equivalent circuit (model 1), while the green line (2) corresponds to the L-R-CPE one (model 2).

Fig. 14 illustrates the Arrhenius plot of electrical conductivity (σ) over the range of 293-473 K, in an argon atmosphere. The experimental data fulfil the linear dependency predicted by the equation:

$$\sigma = \sigma_0 \exp\left[-\frac{E_{act}}{kT}\right] \qquad (11)$$

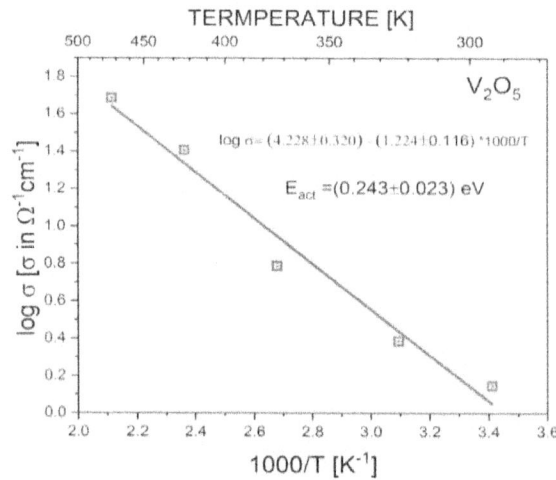

Fig. 14 Temperature dependence of electrical conductivity (Arrhenius plot) of V_2O_5 in an atmosphere consisting of argon/10% H_2.

where the σ_o parameter is independent of temperature, and k denotes the Boltzmann constant. The increase in electrical conductivity with temperature indicates that the material exhibits semiconducting properties in the studied temperature range. The activation energy of electrical conductivity (E_{act}) determined from the slope of the line was (0.243 $\pm$ 0.023) eV. This value is much lower than half of the band gap (E_g) of V_2O_5 – 2.2-2.6 eV [28] predicted for intrinsic electrical conductivity – and it is typical of extrinsic conductivity.

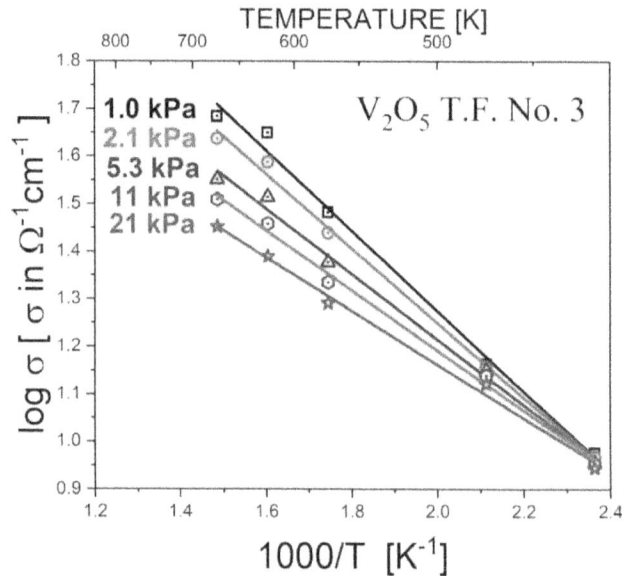

Fig. 15 Temperature dependence of electrical conductivity (Arrhenius plot) of V_2O_5

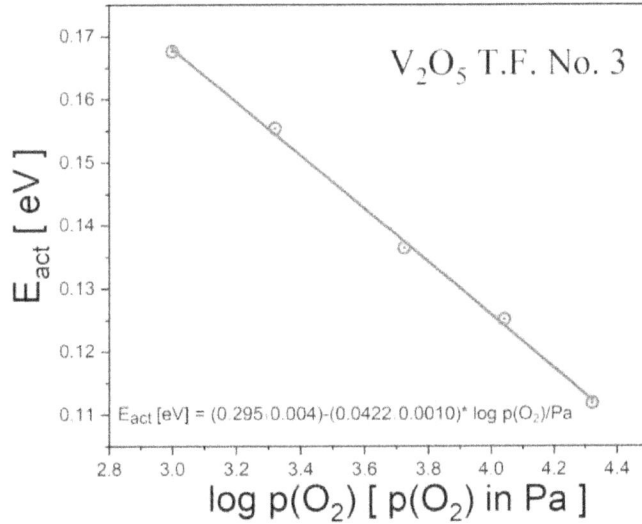

Fig. 16 Activation energy (E_{act}) vs. $p(O_2)$ of a V_2O_5 thin film

The electrical properties of vanadium pentoxide are closely related to its nonstoichimetry (x). Vanadium pentoxide shows a deficit of oxygen in relation to its stoichiometric composition: V_2O_{5-x}. It was reported that the nonstoichiometry x results in the presence of oxygen vacancies and electrons [29]. The determined activation energy (E_{act}) is approximately equal to the enthalpy of electron formation, according to the reaction (12):

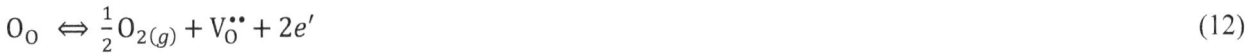

$$O_O \Leftrightarrow \frac{1}{2}O_{2(g)} + V_O^{\bullet\bullet} + 2e' \tag{12}$$

where $V_O^{\bullet\bullet}$ denotes doubly-ionized oxygen vacancies (Kröger-Vink point defect notation).

Fig 15 shows Arrhenius plots of electrical conductivity determined for several oxygen partial pressures and over the temperature range of 423-673 K. As observed for lower temperatures (Fig. 14), the experimental points match the linear relationships well. However, in this case electrical conductivity depends on oxygen partial pressure. The determined activation energy (Fig. 16) is a linear function of $p(O_2)$.

4.3 Electrical conductivity vs. $p(O_2)$

Figs 17 and 18 illustrate the typical experimental results obtained at 623K and 473K for the V_2O_5 thin films, respectively, presented in complex impedance coordinates (Nyquist plots). The relationships of Z'' vs. Z' can be approximated using parts of semicircles. The centers of the semicircles are located below the Z' axis. An increase in the semicircle radius with oxygen partial pressure, $p(O_2)$, is observed.

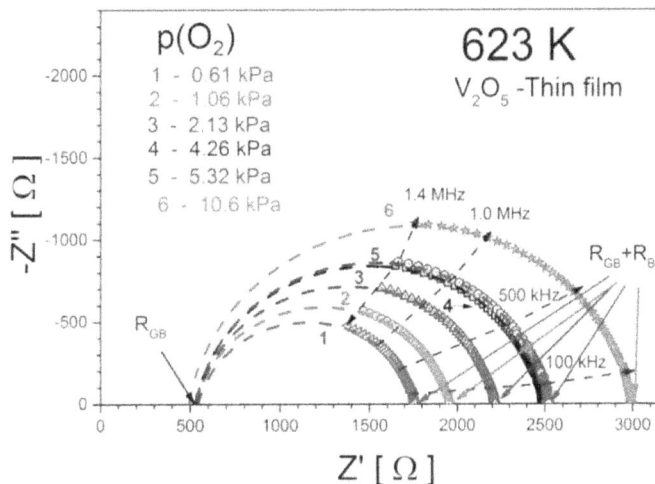

Fig. 17 Z'' vs. Z' for a V_2O_5 thin film

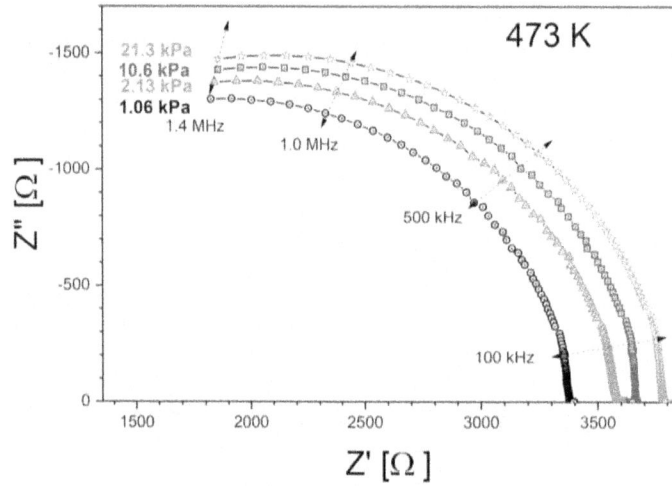

Fig. 18 Z" vs. Z' for a V_2O_5 thin film

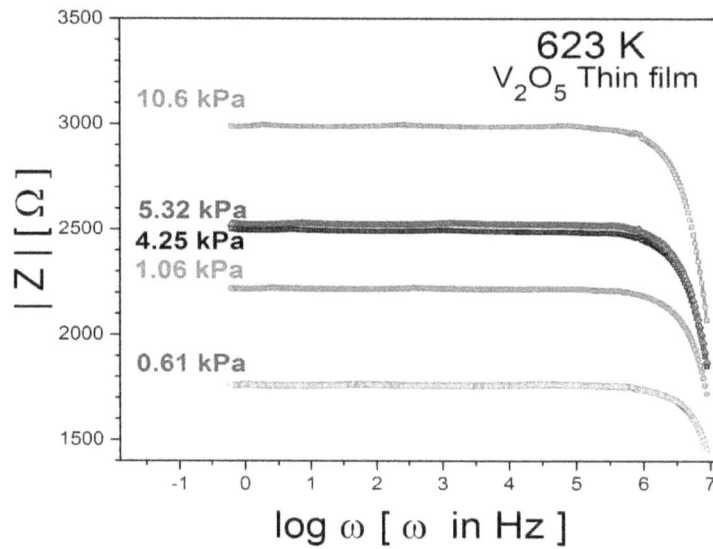

Fig. 19 Modulus of impedance ($|Z|$) as a function of angular frequency (ω) – Bode plot for a V_2O_5 thin film

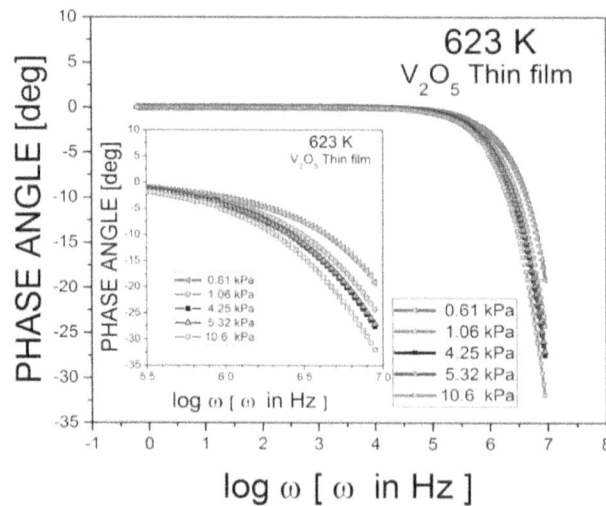

Fig. 20 Phase angle of impedance as a function of angular frequency (ω) – Bode plot for a V_2O_5 thin film

Figs 19 and 20 demonstrate the Bode plots of the modulus of impedance ($|Z|$) and the phase angle, respectively.

Noticeable changes in these variables are observed for frequencies above 50 kHz. The modulus M changes considerably with $p(O_2)$. On the other hand, the phase angle depends on frequency only to a small degree.

4.3.1 Analysis of equivalent circuits

In order to explain the observed phenomena, several equivalent circuits were analyzed. They consisted of a set of resistors, capacitors and inductances with constant parameters (Debye elements) or non-Debye elements such as the constant phase element (CPE) [30]. The impedance behaviour of the equivalent circuit needs to be identical with the behaviour observed for the sample. Moreover, the elements proposed in the equivalent circuit must have a physical meaning. The circuit presented in Fig. 21 is the simplest equivalent circuit that approximated the impedance of the studied samples with good accuracy.

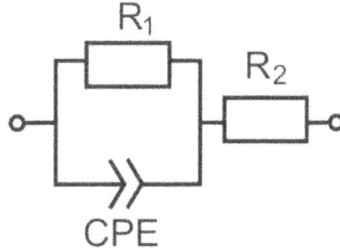

Fig. 21 Equivalent circuit used to analyze the performed impedance measurements.

The impedance of the CPE element (Z_{CPE}) is defined [30] as:

$$Z_{CPE} = A(j\omega)^{-\alpha} \tag{13}$$

where A and α are constants, j is the imaginary unit and ω is angular frequency.

According to the brick wall model, generally accepted to hold true for polycrystalline materials [30], the low-frequency $R_1 = R_B$ and CPE (Fig. 17) is attributed to bulk properties, whereas the high-frequency $R_2 = R_{GB}$ represents grain boundary resistivity. Both R_B and R_{GB} were determined from sections of the semicircles with the Z' axis (Fig. 17).

4.3.2 Defect equilibria

The formation of point defects in V_2O_{5-x} is expressed given by the following reactions (the commonly used Kröger-Vink notation of defects is used throughout this paper):

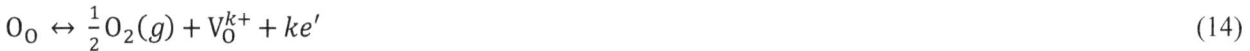

$$O_O \leftrightarrow \frac{1}{2}O_2(g) + V_O^{k+} + ke' \tag{14}$$

where $k = 0$, 1 or 2

$$nil \leftrightarrow e' + h^\bullet \tag{15}$$

The electroneutrality condition requires:

$$n = p + [V_O^\bullet] + 2[V_O^{\bullet\bullet}] \tag{16}$$

where $n = [e']$ and $p = [h^\bullet]$ denote the concentrations of electrons and electron holes, respectively.

When reactions (14) and (15) reach the equilibrium, the law of mass action can be applied. Since the concentrations of point defects in V_2O_{5-x} are rather low, the activities of the defects can be replaced by their concentrations. The equilibrium constant of reaction (14) therefore assumes the following form:

$$K_k(T) = K_k^o \exp\left(-\frac{\Delta H_k}{kT}\right) = [V_O^{k+}]n^k p(O_2)^{\frac{1}{2}} \tag{17}$$

where K_k^o is a constant that accounts for entropy, ΔH_k is the enthalpy of the reaction (14), and k is the Boltzmann constant.

On the other hand, the equilibrium constant of the reaction (15) can be expressed as:

$$K_i = K_i^o \exp(-\frac{E_g}{kT}) \tag{18}$$

where K_i^o is a constant and E_g is the band gap. The equations presented above can be verified by studying some defect relation properties. The most popular is electrical conductivity: $\sigma = \sigma(T, p(O_2))$. The electrical conductivity dependence on oxygen partial pressure can be analysed using the following parameter:

$$m = -\left(\frac{\partial lg\sigma}{\partial lg p(O_2)}\right)_T^{-1} \qquad (19)$$

Assuming that reactions (14) and (15) are in a state of equilibrium, the expected values of parameter m can be estimated from Eqs (14)-(19), assuming four cases of the simplified electroneutrality condition (24). If $k = 0$, then electrical conductivity does not depend on $p(O_2)$ and parameter m tends to infinity; for k equal to 1 or 2 parameter m assumes values of 4 and 6, respectively. These cases are listed in Table 1

Table 1 Expected values of parameter m in the state of equilibrium

No.	Simplified electroneutrality condition	Parameter m
1	$n = [V_O^{\bullet}]$	$m = 4$
2	$n = 2\,[V_O^{\bullet\bullet}]$	$m = 6$
3	$n = [V_O^{\bullet}] + 2\,[V_O^{\bullet\bullet}]$	$4 < m < 6$
4	$n = p$	∞

The results of log σ_B vs. log $p(O_2)$ at 623, 473 and 423 K are shown in Fig. 22. At an elevated temperature (623 K), parameter m assumes a value that corresponds to the co-existence of both singly ($V_O^{\bullet}$) and doubly ($V_O^{\bullet\bullet}$) ionized oxygen vacancies. This fact indicates that the entire thin film material interacts with the gas atmosphere. This conclusion explains the relatively high response times of the V_2O_5 thin film sensors, as reported in Ref. [29]. The dependence of grain boundary electrical conductivity on oxygen partial pressure observed at 623 K is very different (Fig. 22) – the negative values of m are specific to p-type electrical conductivity. The negative value of parameter m in the case of the σ_{GB} at 623 K (Fig. 22) can be due to phenomena that occur at grain boundaries and the surface, such as chemisorption or segregation. On the other hand, at lower temperatures, i.e. 423 and 473 K, the parameter m for bulk electrical conductivity (σ, Fig. 22) assumes values of 43.5 and 19.8 K, respectively, whereas for σ_{GB} and at 423 K it is equal to 8.9 (Fig. 23). Such results suggest that either reaction (14) in the bulk is not in a state of equilibrium, or that there is a substantial contribution of neutral oxygen vacancies ($k = 0$). It can be surmised that at lower temperatures (Fig. 22) the interaction of oxygen with the V_2O_5 thin film is limited to the solid surface and grain boundaries (i.e. paths of fast diffusion), and it may be interpreted in terms of phenomena that occur on the surface and in grain boundary regions.

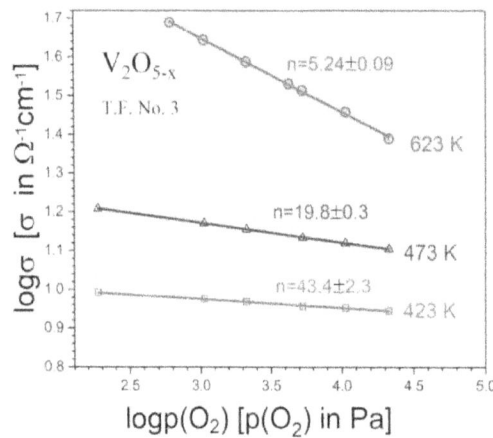

Fig. 22 Electrical conductivity (σ) as a function of oxygen partial pressure ($p(O_2)$) at 423, 473 and 623 K.

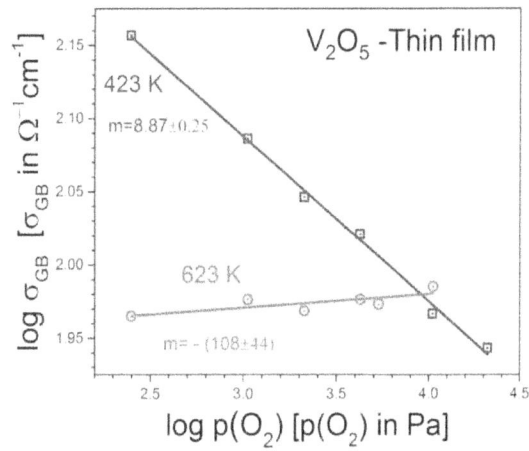

Fig. 23 Grain boundary electrical conductivity (σ_{GB}) as a function of oxygen partial pressure ($p(O_2)$) at 423 and 623 K.

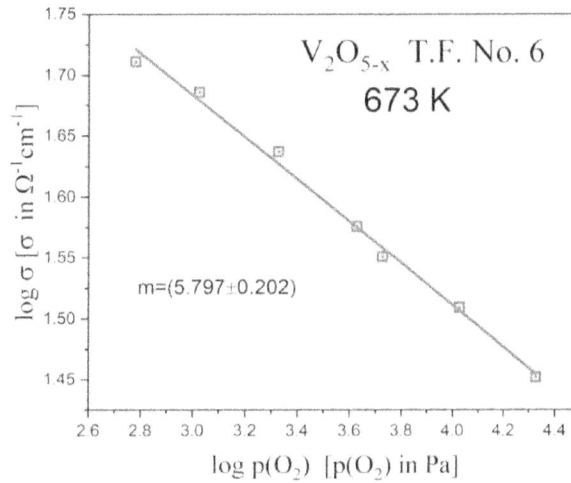

Fig. 24 Dependence of electrical conductivity (σ) on oxygen partial pressure ($p(O_2)$) (V_2O_5 T.F.)

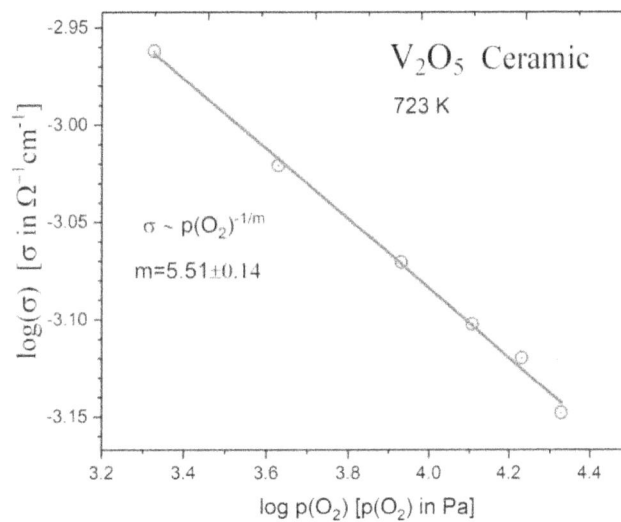

Fig. 25 Dependence of electrical conductivity (σ) on oxygen partial pressure ($p(O_2)$) (ceramic V_2O_5 sample)

Figs 24 and 25 present the dependences of log σ versus log $p(O_2)$ at 673 K – for a thin film, and at 723 K – for ceramic vanadium pentoxide. The experimentally determined parameter m can be plained assuming a deviation from stoichiometry towards a deficit of oxygen in the studied materials: V_2O_{5-x}. The nonstoichiometry x results from reaction

95

(14).

If expressed via the concentrations of point defects in the oxygen sub-lattice fraction, the nonstoichiometry x in V_2O_{5-x} is:

$$x = [V_O^x] + [V_O^{\bullet}] + [V_O^{\bullet\bullet}] \tag{20}$$

According to Fig. 22, at lower temperatures (423 and 473 K) the corresponding values of parameter m can indicate that most oxygen vacancies are un-ionized.

On the other hand, at 623 K and 673 K the parameter m assumes values between 4 and 6 ($4 < m < 6$). The data in Table 1 indicate that singly and doubly ionized oxygen vacancies are present at the same time. Electrical conductivity should therefore obey the following relationships:

experimental:

$$\sigma_{exper.} = Cp(O_2)^{-\frac{1}{m}} \tag{21}$$

and theoretical:

$$\sigma_{theor.} = e\,n\mu_n = e\mu_n * \{[V_O^{\bullet}] + 2[V_O^{\bullet\bullet}]\} = Bp(O_2)^{-\frac{1}{4}} + 2Ap(O_2)^{-\frac{1}{6}} \tag{22}$$

where e is elementary charge, μ_n is the mobility of electrons and A, B and C are the parameters independent of $p(O_2)$. Parameters B and A were calculated using the least squares method:

$$Z = \sum_{i=1}^{i=N} [\sigma_i - Bp(O_2)_i^{-\frac{1}{4}} - 2A(O_2)_i^{-\frac{1}{6}}]^2 \tag{23}$$

where N is the number of experimental points (σ_i, $p(O_2)_i$)

From the minimum criterion of $Z = Z_{min}$:

$$\frac{\partial Z}{\partial A} = 0 \; and \; \frac{\partial Z}{\partial B} = 0 \tag{24}$$

a system of two linear equations for unknown values of parameters A and B is obtained.

As per Eq. (22), the determined values of parameters A and B are indicative of the mutual contribution of doubly and singly ionized oxygen vacancies in electrical conductivity. The mean contributions of the concentrations of doubly ($V_O^{\bullet\bullet}$) and singly ($V_O^{\bullet}$) ionized oxygen vacancies to n-type electronic conductivity can be expressed as $\frac{A}{A+B}$ and $\frac{B}{A+B}$.

The results of calculations are listed in Table 2 and presented in Figs 26, 27 and 28 for Series 3 and Series 6 V_2O_5 thin films and the V_2O_5 ceramic sample, respectively.

Table 2 Mean contribution of ionic point defects to electrical conductivity

Specimen	T [K]	$V_O^{\bullet\bullet}$ [%]	$V_O^{\bullet}$ [%]
V_2O_5 Thin Film (T.F.)	623	58.8	41.2
V_2O_5 Thin Film (T.F.)	673	87.4	12.6
V_2O_5 Ceramic	723	57.9	42.1

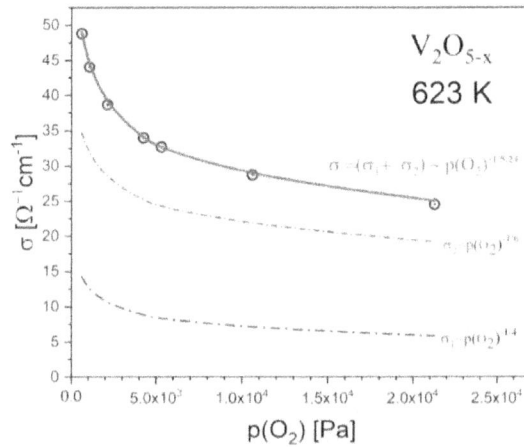

Fig. 26 Electrical conductivity (σ) as a function of oxygen partial pressure ($p(O_2)$)) at 623 K (V_2O_5 T.F.)

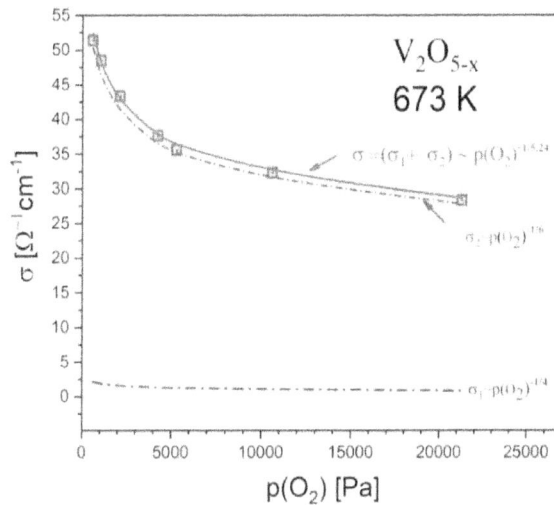

Fig. 27 Electrical conductivity (σ) as a function of oxygen partial pressure ($p(O_2)$)) at 673 K (V_2O_5 T.F.)

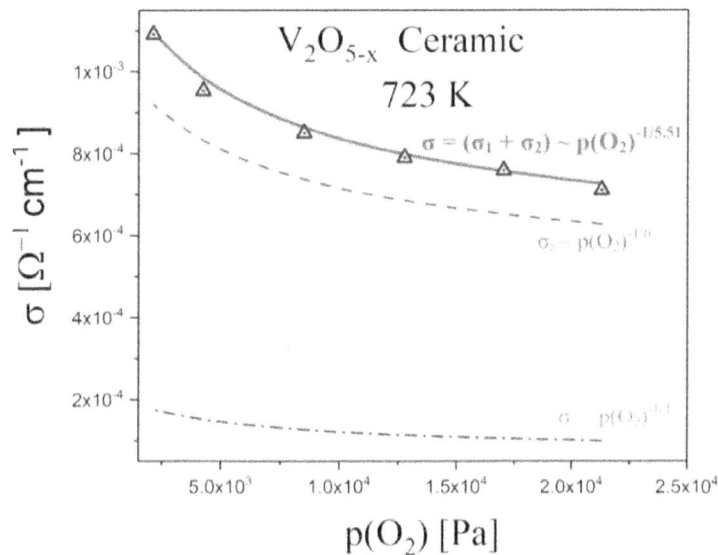

Fig. 28 Electrical conductivity (σ) as a function of oxygen partial pressure ($p(O_2)$)) at 723 K (V_2O_5 ceramic)

4.4 Electrical conductivity vs. time

The formation of defects described in section 4.3 depends mainly on thermodynamic and kinetic factors. The former of these indicates the most favorable state which can be reached in the material. It is determined by the thermodynamic functions of the defect formation reactions [32]. The second determines whether a state of equilibrium can actually be achieved in a finite time. It is controlled by the slowest step of the process, usually diffusion. The diffusion process is characterized by a diffusion coefficient.

Most of the data concerning the transport of mass in semiconducting oxides have been collected from ionic electrical conductivity results. The diffusion coefficient can then be estimated based on the Nernst-Einstein relationship.

The purpose of this work was to investigate chemical diffusion in vanadium pentoxide using a transient electrical conductivity method.

Equilibrium states of metal oxide/oxygen systems are determined by temperature and oxygen partial pressure. At equilibrium, the concentrations of point defects and the related properties such as deviation from stoichiometry (x) and electrical conductivity (σ) are well-defined. When either temperature or $p(O_2)$ suddenly changes, the system tends to reach a new equilibrium state, which is initially imposed at the oxide/gas phase boundary. The equilibration involves the propagation of defects from the boundary and into the bulk phase. This process is rate-controlled by chemical diffusion and can be monitored by conducting measurements of oxide properties sensitive to defect concentration, such as oxide mass or electrical conductivity.

The purpose of this work was to investigate chemical diffusion in vanadium pentoxide using a transient electrical conductivity method.

The equilibration kinetics is usually expressed as the dependence of equilibration degree (γ) on time [33, 34]:

$$\gamma = \gamma_x = \frac{x_t - x_0}{x_\infty - x_0} = \frac{\Delta m_t}{\Delta m_\infty} \tag{25}$$

where x is the deviation from stoichiometry in V_2O_{5-x}, Δm is mass change and subscripts t, 0 and ∞ correspond to time, the initial state, and the final state, respectively.

As follows from definition (25), the re-equilibration kinetics may be monitored by measuring a property which is directly correlated to the change in non-stoichiometry (Δx) accompanying the equilibration process, such as mass change (Δm). However, in the case of thin films the expected mass changes can be below the sensitivity level of the termogravimetric method. Hence, monitoring the equilibration process requires the measurement of a defect-related property which is sensitive to changes in non-stoichiometry. Electrical conductivity has most frequently been used for this purpose [35, 36]. This method was successfully applied with oxide materials that exhibit extrinsic electronic conductivity. In such cases, there is a simple relationship between non-stoichiometry and electrical conductivity. Consequently, the equilibration degree (γ_x) is equal to γ_σ, which is expressed as:

$$\gamma = \gamma_\sigma = \frac{\sigma_t - \sigma_0}{\sigma_\infty - \sigma_0} \tag{26}$$

Based on the experimentally determined relationship of σ vs. time and Eq. (26), the equilibration degree (γ) was determined as a function of time. Figs 29 and 30 illustrate the typical dependences for a Series 3 V_2O_5 T.F. at a temperature of 623 K; these involve changes in $p(O_2)$ – from: 4.2 kPa to 10.5 kPa (Fig. 29) and from 21 kPa to 190 Pa (Fig. 30).

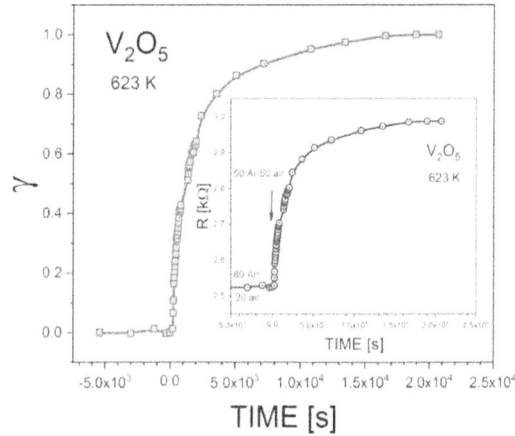

Fig. 29 Equilibration degree (γ) and resistance (R) vs. time equilibration corresponding to a change in $p(O_2)$ from 4.2 kPa to 10.5 kPa (623K)

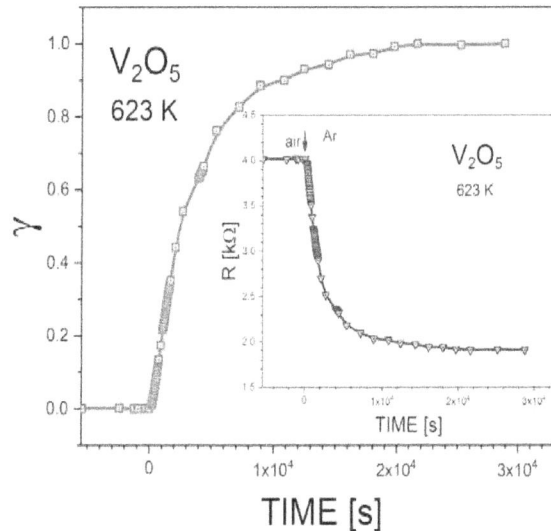

Fig. 30 Equilibration degree (γ) and resistance (R) vs. time equilibration corresponding to a change in $p(O_2)$ from 21 kPa to 190 Pa (623K).

The presented dependence of γ vs. time may be used to determine the chemical diffusion coefficient – D_{chem}. It may be calculated by solving Fick's second law under appropriate initial and boundary conditions:

a) **Initial condition**

The equilibrium state for vanadium pentoxide/gas oxygen is determined by two parameters: temperature and oxygen partial pressure. This state may be observed by monitoring electrical conductivity. When electrical conductivity is independent of time, it can be assumed that the gas/solid system is in equilibrium. In the equilibrium state, the deviation from stoichiometry in V_2O_{5-x} and other related parameters such as the concentration of point defects are constant for the entire sample. Hence, the following initial condition can be formulated for a brick-shaped specimen with the dimensions $a \times b \times c$:

$$x(\tau, \chi, \xi, t) = \text{constant} \tag{27}$$

where t is time, τ, χ, ξ represent space coordinates, and $0 < \tau < a$, $0 < \chi < b$, $0 < \xi < c$.

b) **Boundary condition**

When $p(O_2)$ changes abruptly, the system tends to reach a new equilibrium state. This process involves two steps. The first is either the adsorption or desorption of oxygen molecules at the film's surface (a surface

99

reaction). The second stage is the propagation of oxygen ions into the bulk phase. It is assumed that the first step is quick compared to the second. This means that the new equilibrium state is initially imposed at the gas/solid interface, and parameters such as the x in the interface are consistent with the new equilibrium. Hence, the following can be assumed to be true on the film's surface after the change in $p(O_2)$:

$$x(surface, t) = constant \text{ for } t > 0 \tag{28}$$

Taking into consideration the initial (27) and boundary (28) conditions, the solution of Fick's second law is [31-37]:

$$\gamma = 1 - \frac{8}{\pi^2} \sum_{n=0}^{\infty} \exp\left[-\frac{(2n+1)^2 \pi^2 D_{chem} \, t}{d^2}\right] \tag{29}$$

where d is film thickness.

Eq. (29) can be simplified for two γ regions:

- At lower values of the degree of equilibration ($\gamma < 0.5$), the diffusion in a sample with a rectangular shape may be approximated in terms of the diffusion in a semi-infinite medium. This causes the following equation to replace Eq. (29):

$$\gamma = \left(\frac{D_{chem} t}{\pi}\right)^{\frac{1}{2}} \frac{2S}{V} \tag{30}$$

where S and V denote the surface and volume of the sample, respectively.

Equation (30) is called the 'parabolic equation of the solution of Fick's second law.

- At a higher degree of equilibration ($\gamma > 0.7$), the first term ($n = 0$) in the sum in Eq. (29) is large compared to the terms that follow it ($1 \leq n < \infty$), which is why the following is true:

$$\log(1 - \gamma) = \log\frac{8}{\pi^2} - \frac{\pi^2 D_{chem} t}{4d^2 \ln 10} \tag{31}$$

Eq. (31) is called the 'logarithmic equation of the solution of Fick's second law.

Figs 31 and 32 present the typical parabolic and logarithmic plots of the kinetic data shown in Fig. 5.35. The experimental data fit those theoretically predicted by equations (30) and (31) well.

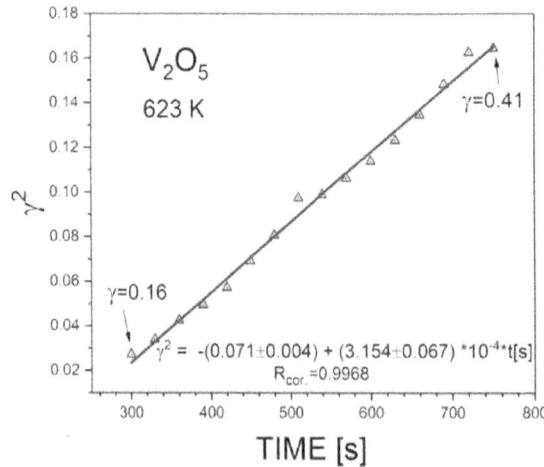

Fig. 31 Parabolic plot of an equilibration isotherm at 623 K, corresponding to a change in $p(O_2)$ from 4.2 kPa to 10.5 kPa

100

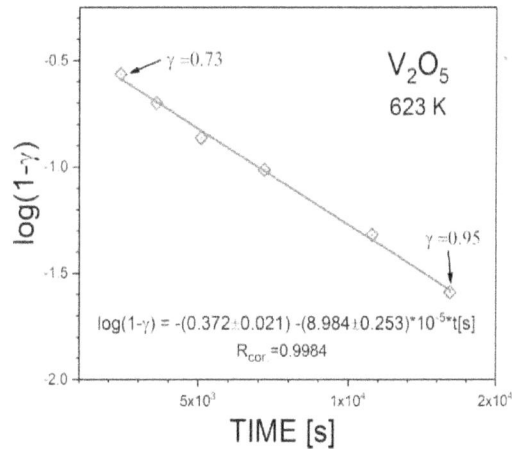

Fig. 32 Logarithmic plot of an equilibration isotherm at 623 K, corresponding to change in $p(O_2)$ from 4.2 kPa to 10.5 kPa.

Figs 33 and 34 illustrate the Arrhenius plots of chemical diffusion coefficients determined from the parabolic ($D_{chem, par}$) and logarithmic ($D_{chem, log}$) equations, respectively.

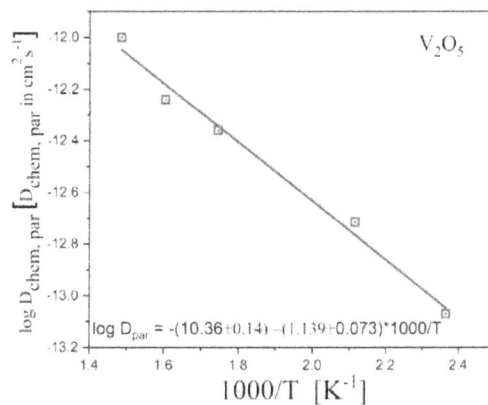

Fig. 33 Arrhenius plot of the chemical diffusion coefficient ($D_{chem,par}$) determined from the parabolic equation

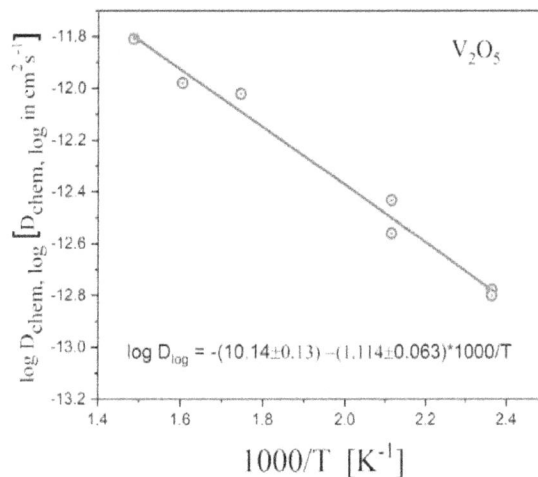

Fig. 34 Arrhenius plot of the chemical diffusion coefficient ($D_{chem,log}$) determined from the logarithmic equation

A comparison of the two sets ($D_{chem,par}$ and $D_{chem,log}$) of data is presented in Fig. 35. It was determined via least squares fitting found that both sets of D_{chem} obey the Arrhenius rule:

$$D_{chem} = D^0_{chem} \exp[-\frac{E_{act}}{kT}] \tag{32}$$

where D^0_{chem} is a parameter independent of temperature and E_{act} is the activation energy of diffusion.

The determined activation energy values are practically identical for $D_{chem,par}$ and $D_{chem,log}$. The observed differences in D^0_{chem} values determined from parabolic and logarithmic equations may result from approximations of both equations as well as from the uncertainty of measurement results.

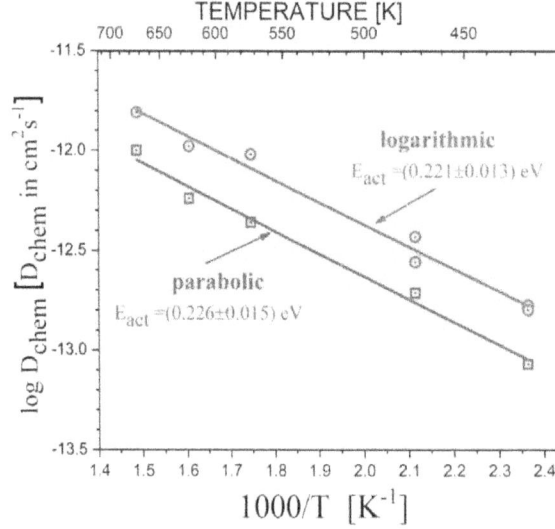

Fig. 35 Comparison of the chemical diffusion coefficients determined from parabolic and logarithmic equations.

Taking into account the mean values of both D_{chem} sets, the determined chemical diffusion coefficient has the following value:

$$D_{chem} = (5.6 \pm 1.8) \cdot 10^{-11} \cdot \exp\left[-\frac{(0.224 \pm 0.014)eV}{kT}\right] [cm^2 s^{-1}] \tag{33}$$

The physical meaning of the chemical diffusion coefficient is expressed by the following relation [38]:

$$\delta = \sqrt{D_{chem} t(\delta)} \tag{34}$$

where δ is the penetration depth of diffusion and $t(\delta)$ is the time needed by the diffusive species to achieve depth δ. In particular, the time equilibration of the entire film assumes the following value:

$$t(d) = \frac{d^2}{D_{chem}} \tag{35}$$

The equilibration time calculated from Eq. (35) for the 200 nm thin film at 423 K is ca. 50 min.

5. CONCLUSIONS

Defect structure of VO_2 thin films

In the present paper, the main characteristics of vanadium oxide thin films were discussed. The as-deposited films were amorphous. The nominal chemical composition determined from the RBS experiments was $V_{1.000}O_{2.030}$. No diffusion between the material of the film and the applied substrate was observed. Hydrogen treatment did not affect the accumulation of this element in the film material.

Electrical properties & defect structure of V_2O_5 thin films

Electrical conductivity vs. T

The electrical properties of V_2O_5 were investigated by analyzing the complex impedance spectra at frequencies ranging from 0.1 Hz to 1.4 MHz and as a function of temperature over the range of 290-773 K. Two entirely different impedance spectra were obtained. At temperatures below 520 K, the recorded impedance spectra corresponded to an equivalent circuit composed of a resistor and a non-Debye constant phase element connected in parallel. In this temperature range the material exhibits n-type extrinsic conductivity. The activation energy of electrical conductivity was 0.243 ± 0.023 eV in an Ar/10% H_2 gas atmosphere. Linear changes in E_{act} versus $\log p(O_2)$ for 1 kPa $< p(O_2) <$ 21 kPa were observed. At 528 K, an abrupt change in resistivity was observed. This phenomenon was interpreted as a metal-insulator transition (MIT). Above 528 K, the recorded impedance spectra can be interpreted using an equivalent circuit composed of a resistor, an inductor, and a CPE element connected in series. The CPE element has only a minor effect on impedance spectra. Above 528 K, the studied sample exhibited metallic properties.

Defect structure of V_2O_5

The point defect structure of the V_2O_5 thin films was investigated by analyzing the complex impedance spectra at frequencies ranging from 0.1 Hz to 1.4 MHz and as a function of oxygen partial pressures varying from 600 Pa to 21 kPa. The recorded impedance spectra were highly consistent with the equivalent circuit composed of two resistors and a constant phase element. A distinct effect of the temperature was observed. At elevated temperatures close to 620 K, the entire thin film was found to interact with oxygen. Both singly and doubly ionized oxygen vacancies and electrons were the product of these interactions. At lower temperatures close to 420 K, gas-solid interactions were limited only to the surface of the film. The observed results have major implications with regard to V_2O_5 thin film resistive-type gas sensors.

In this chapter/section the electrical properties of V_2O_5 were investigated by analyzing the complex impedance spectra at frequencies ranging from 0.1 Hz to 1.4 MHz and as a function of temperature over the range of 290-773 K. Two entirely impedance spectra were obtained. At temperatures below 520 K impedance spectra correspond to the equivalent circuit composed from a resistor, R, and non-Debye constant phase element, CPE, connected parallel. In this temperature range the material exhibits n-type extrinsic conductivity. Activation energy of the electrical conductivity was $E_{act} = (0.243 \pm 0.023)$ eV at Ar/10% H_2 gas atmosphere. The linear changes of E_{act} versus $\log p(O_2)$ within 1 kPa $< p(O_2) <$ 21 kPa was observed. At 528 K the abrupt change of resistivity is observed. This phenomenon was interpreted as metal-insulator transition (MIT). Above 528 K the impedance spectra can be interpreted using an equivalent circuit composed from resistor R, inductor, L and CPE element, connected in Series. The CPE element has only minor effect on the impedance spectra. Above 528 K studied sample has metallic property.

Chemical Diffusion

To the best of the author's knowledge, this work is the first study of chemical diffusion in vanadium pentoxide. Measurements of transient electrical conductivity made it possible to determine the chemical diffusion coefficients based on the re-equilibration kinetics of vanadium pentoxide thin film in the temperature range of 423-673 K and for the following oxygen partial pressure range: 190 Pa $< p(O_2) <$ 210 kPa. The determined dependence of the chemical diffusion coefficient on temperature can be expressed by the following equation:

$$D_{chem} = (5.6 \pm 1.8) \cdot 10^{-11} \cdot \exp\left[-\frac{(0.224 \pm 0.014)eV}{kT}\right] [cm^2 s^{-1}] \tag{36}$$

REFERENCES

1. F.J. Morin 1959. *Oxides which show a metal-to-insulator (MIT) transition at the Neel temperature*, Phys. Rev. Lett. 3 (1959) 34-36

2. K. Schneider, *Vanadium oxides – properties and applications, Part III Metal-Insulator Transition (MIT) in vanadium oxides*, This issue)

3. K.Schneider, *Vanadium oxides – properties and applications, Part IV Thin films: preparation, properties, applications,* this issue.

4. B Sass, C Tusche, W Felsch, N Quaas, A Weismann and M Wenderoth, *Structural and electronic properties of epitaxial* V_2O_3 *thin films*, J. Phys. Condens.Mater. 16 (2004) 77-87.

5. V. Simic-Milosevic, N. Nilius, H.-P. Rust, and H.-J. Freund, *Local band gap modulations in non-stoichiometric* V_2O_3 *films probed by scanning tunneling spectroscopy,* Phys. Rev. B 77 (2008) 125112, 5

6. D. Wickramaratne, N. Bernstein, and I. I. Mazin, *The role of defects in the metal-insulator transition in* VO_2 *and* V_2O_3, Phys. Reb. B , accepted 6 May 2019.

7. D.B. McWhan, T.M. Rice, J.P. Remeika, *Mott transition in Cr-doped* V_2O_3, Phys. Rev.Lett.23 (1969) 1384

8. M. Wakihara, T. Katsura, *Thermodynamic properties of the* V_2O_3-V_4O_7 *system at temperatures from* 1400° *to* 1700° *K*, J.Phys.Chem. Soc. Japan 1 (1970) 363-366.

9. H. Kim, N. Charipar, M. Osofsky, S. B. Qadri, and A. Pique, *Optimization of the semiconductor-metal transition in* VO_2 *epitaxial thin films as a function of oxygen growth pressure.* Appl.Phys.Lett. 104 (2014) 081913, 5

10. C. Blaauw, F. Leenthouts, F. van der Woude, G.A. Sawatzky, *The phase transition in* VO_2, J.Phys.C8 (1975) 459

11. Q. Yu, W. Li, J. Liang, Z. Duan, Z.Hu, J.Liu, H. Chen, J.Chu, *Oxygen pressure manipulations on the metal–insulator transition characteristics of highly* (011)-*oriented vanadium dioxide films grown by magnetron sputtering,* J. Phys. D: Appl.Phys. 46 (2013) 055310, 7

12. H. Kim, J. Jeong, N. Aetukuri, T. Graf, T.D. Schladt, M.G. Samant, S.S.P. Parkin , *Supression of metal-insulator transition in* VO_2 *by electric field-induced oxygen vacancy formation,* Science 339 (2013) 1402-140

13. K. Appavoo, D.Y. Lei, Y. Sonnerfraud, B. Wang, S.T. Pentelides, S.A. Maier, R.F. Haglund Jr, *Role of defects in the phase transition of* VO_2 *nanoparticles probed by plasmon resonance spectroscopy,* Nano. Lett. 12 (2012) 780-786

14. Y. Cui, B. Liu, L. Chen, H. Luo, Y. Gao, *Formation energies of intrinsic point defects in monoclinic* VO_2 *studied by first-principles calculations,* AIP Adv. 6 (2016) 105301, 9

15. D. Wickramaratne,, N. Bernstein, I.I. Mazin, *The role of defects in the metal-insulator transition in* VO_2 *and* V_2O_3, Phys.Rev. B, accepted (2019) 9.

16. J. Wei, H. Li, W. Guo, A.H. Nevidomskyy, D. Natelson, *Hydrogen stabilization of metallic vanadium dioxide in single crystal nanobeams,* Nature Nanotecholog. 7 (2012) 357-362.

17. K.H. Warnick, B. Wang, S.T. Pentelides, *Hydrogen dynamics and metallic phase stabilization in* VO_2, Appl.Phys.Lett. 104 (2014) 101913.

18. P. Pasierb, M. Rekas, *High-temperature electrochemical hydrogen pumps and separators,* Intern. J. Electrochem. 2011 (2011) 905901, 10.

19. K. Schneider, K. Zakrzewska, Z. Tarnawski, K. Drogowska, N-T. H. Kim-Ngan, VO_x *thin films deposited by reactive rf sputtering,* Ceramics v. 115 (2013) 305-314

20. N.T.H.Kim-Ngan, A.G.Balogh, J.D.Meyer et al., *Thermal and irradiation induced interdiffusion in magnetite thin films grown on magnesium oxide (0 0 1) substrates,"* Surface Science, 603 (2009) 1175-118.

21. M.Mayer, AIP Conference Proceedings, vol. 475, 1999, *SIMNRA Simulation Program for the Analysis of m NRA, RBS and ERDA)* developed by M.Mayerhttp://www.rzg.mpg.de/~mam/., 541-544.

22. Y. Yang, D. Ki M. Yang, P. Schmuki, *Vertically aligned mixed* V_2O_5–TiO_2 *nanotube arrays for supercapacitor applications,* Chem. Commun. 47 (27) (2011) 7746-7748.

23. H.Iwahara, T.esaka, H.Uchida, N.Maeda, *Proton conduction in sintered oxides and its application to steam electrolysis for hydrogen production,* Solid State Ionics 3.4 (1981) 359-363.

24. Y. Cui, B. Liu, L. Chen, H. Luo, Y. Gao, *Formation energies of intrinsic point defects in monoclinic* VO_2 *studied by first-principles calculations,* AIP Advances 6 (2016) 15301.

25. S.Fan, L. Fan, Q.Li, J. Liu, B. Ye, *The identification of defect structures for oxygen pressure dependent* VO_2 *crystal films*, App. Surf.Sci. 321 (2014) 464-468.

26. R.D. Shannon Revised effective ionic radii and systematic studies of interatomic distances in halides and chalcogenides, Acta Cryst. A32 (1976) 75767.

27. S.J. Petel, V. Kheraj, Determination of refractive index and thickness of thin-film from reflectivity spectrum using genetic algorithm, AIP Conference Proceedings 1536, 509 (2013); https://doi.org/10.1063/1.4810324.

28. K.Schneider, *Vanadium oxides – properties and applications, PartV. Electronic structure*, this issue.

29. Kofstad P. in: Nonstoichiometry, Diffusion and electrical in binary metal oxides, Wiley, Intersci., NY, London, Sydney, Tokyo, 1972.

30. J.R. Macdonald, D.R. Franceschettu, Impedance spectroscopy, edited by J.R. Macdonald (John Wiley & Sons), New York, 84 (1987).

31. K. Schneider, M. Lubecka, A. Czapla, V_2O_5 *thin films for gas sensor applications*, Sensors and Actuators B Chem. 263 (2016) 970-977.

32. K. Schneider, *Vanadium oxides – properties and applications.* II *Vanadium oxide electronics,* this issue.

33. J. Cranc, in: Mathematics in diffusion, Oxford Press, London 1956.

34. J.B. Price, J.B. Wagner Jr, *Crystalline manganeous oxide*, J. Electrochem. Soc. 117 (1970) 242-247.

35. P.E. Childs, L.W. Laub, J.B. Wagner Jr, *Chemical diffusion in non-stoichiometric compounds,* Proc.Br. Ceram. Soc. 19 (1971) 29-38.

36. P. Pasierb, S. Komornicki, M. Rekas, *Comparison of the chemical diffusion of undoped and Nb-doped* $SrTiO_3$, J.Phys.Chem. Solids 60 (1999) 1835-1844.

37. M. Radecka, P. Pasierb, K. Zakrzewska, M. Rekas, *Transport properties of (Sn,Ti)O_2 polycrystalline ceramics and thin films*, Solid State Ionics 119 (1999) 43-48.

38. N.B. Hannay in Solid-state chemistry, Prentice Hall Inc, Englewood Cliffs, New Jersey, USA 1967, 131.